Pittsburgh Antique Radio Society Publications
David W. Kraeuter, General Editor

1. *A Bibliography of Frank Conrad,* (Third Edition), 2007.
2. *The U.S. Patents of Reginald A. Fessenden,* (Second Edition), 2007.
3. *A New Bibliography of Reginald A. Fessenden,* (Second Edition), 2007.
4. *The U.S. Patents of John H. Hammond, Jr.,* (Second Edition), 2007.
5. *An Interview with Harold Beverage,* Richard Brewster, (Second Edition), 2007.
6. *Radio and Television Reminiscences: Raymond M. Bell in the Pittsburgh Oscillator,* (Third Edition), 2007.
7. *Electronic Essays,* David W. Kraeuter, (Sixth Edition), 2012.
8. *The U.S. Patents of Harold S. Black, Jack S. Kilby and Robert N. Noyce,* (Second Edition), 2007.
9. *The U.S. Patents of Stuart W. Seeley* (with a bibliography of Seeley's writings), (Second Edition), 2007.
10. *Ten Patents from Radio History,* David W. Kraeuter, 2007.
11. *Frank Conrad's Radio Patents: The Complete Texts,* (Second Edition), 2007.
12. *Electronic Reviews: Hundreds of Thoughts on 100 Books,* David W. Kraeuter, 2008.
13. *The 3 Strikes Camp Stories,* Karl Laurin, 2008.
14. *Vintage Radio Redux,* (Second Edition), Karl Laurin, 2012.
15. *A Radio Patent Chronology,* David W. Kraeuter, 2009.
16. *25 Years of Electronic Reviews,* David W. Kraeuter, 2011, 2014.
17. *The Pittsburgh Antique Radio Society at 25,* 2011.
18. *Sketches from a Life in Electronics,* Laten Fetters, 2012.
19. *Mr. Johnson Answers,* William (Bill) Johnson, Jr., 2012.
20. *An Oscillator Reader,* 2013.
21. *Meetings, Contests, Winners,* 2014.
22. *An Oscillator Reader II,* 2014.
23. *The Pittsburgh Antique Radio Society at 30,* 2016.
24. *1 2 3 4 5 6: A David Kraeuter Sampler,* 2017.
25. *The* At Home *Series,* Rex Kraeuter, 2018.
26. *The Pittsburgh Antique Radio Society at 35,* 2021.
27. *35 Years of Electronic Reviews,* David W. Kraeuter, 2022.

 All books available at *www.lulu.com*

35 Years of Electronic Reviews

David W. Kraeuter

Pittsburgh Antique Radio Society Publication 27

Available from: *www.lulu.com*

Preface

I started writing book reviews in grade school. Today they make amusing reading. But don't worry. All the reviews here were written in the last 35 years. They are of books in the field of electronics in general, and in radio history specifically.

The reviews range in time as far back as the imaginary compass of Famianus Strada in 17th century Rome and as recently as the publication of Ludwell Sibley's *Tube Lore II.* Subjects range from the fine esoterica of collecting, as in Eric Wenaas's beautiful Radiola book, to the mundane but occasionally vital detail of Alfred Ghirardi's *Radio Troubleshooter's Handbook.*

A note on book availability appears on page 217 and an author index begins on page 219. The author himself appears on the front cover and on page 224.

I thank the *AWA Journal, The OTB, The Old Timer's Bulletin* and the *Pittsburgh Oscillator* for originally printing the reviews Citations are provided for those publications. Note that initial articles *a, an* and *the* have been deleted in the book titles to facilitate alphabetization.

Lastly, you, dear reader, I suspect have arrived at this volume by a somewhat vague or peculiar route. You may be advised to look elsewhere. For, as Montaigne once famously wrote, *Que sais-je?*

David W. Kraeuter
Washington, Pennsylvania
January 18, 2022

This book is dedicated to

Frances Watona

and to the memory of

Richard Kraeuter, 1939-2010

Artwork credits: front cover photo by Robert Sullivan, about 1992; title page drawing, DWK.

ISBN: 978-1-716-38035-8

A. Atwater Kent: The Man, the Manufacturer and His Radios
Ralph Williams, Sonoran Publishing, LLC, 108 pages, 2002.
OTB, v44 n2, May 2003, p. 58.

Another beauty from Sonoran, this book is largely an adaptation from a series of articles Williams wrote for *Radio Age* beginning in 1976. The original articles are no longer in print, so this makes a handy compilation of material not otherwise available. The book includes a 12-page essay, "Worcester Polytechnic Institute's Forgotten Millionaire" by John P. Wolkonowicz.

About half the book is devoted to the breadboards, and many of these have circuit diagrams or monochrome photographs of the top and bottom of the boards. There are also three pages of color photographs, mostly of AK advertising items.

Williams obviously knew his subject thoroughly, even in the "dark ages" of radio history (the 1970s). So it's not surprising that he doesn't hesitate to state his opinions about AK matters for which there may be little or no direct evidence today. See, for example, his perceptive discussion on page 20 of Kent's reluctance to embrace the regenerative or reflex circuits in his radios.

Another typical insight by Williams appears on page 19 in a paragraph about the low audio output of Model 10 breadboards:

> *In the days when AK instruments were new, radio was so novel, so uncommon, and so expensive in both money and time, that it was accorded the quietness of surroundings that made hearing by a group possible, even with the limited volume available.*

The all-AK *AWA Review*, volume 12, 1999 can be considered a companion volume, since it is also by this master AK historian, Ralph Williams. AK enthusiasts will want to own both.

Alexanderson: Pioneer in American Electrical Engineering
James E. Brittain, Johns Hopkins University Press, 381 pages,
  1992.
AWA Journal, v48 n1, January 2007, p. 17.

   Brittain has produced a comprehensive biography of his subject. It
is the only full-length biography of Alexanderson to appear so far.
Brittain's academic background was appropriate for such a
biography–his 1970 Ph.D. dissertation was titled *B. A. Behrend and
the Beginnings of Electrical Engineering, 1870-1920.*
   He covers Alexanderson's extensive career at General Electric in
great detail. Alexanderson's interests ranged widely, but
encompassed two related fields–electronics and power generation
and transmission. He often stated that each field fertilized the other
in his thinking and work–witness his radio alternator, an example of
which is still in existence in Grimeton, Sweden.
   My only quibble with the book is that the text is plodding and
unexciting. What, for example, are we to make of a sentence like
this: "Remarking that cooperative activity among those responsible
for design, research, manufacture, and sales would be required in
order to regain a position of leadership, he observed that it was the
aggregate momentum of all these sectors that constituted the great
strength of the company."? It is clear, accurate, grammatically
correct–and lifeless. During his 40+ years with General Electric and
many years with RCA, Alexanderson wrote many annual reports and
hundreds of memos. Brittain seems to have read them all and
summarized many of them.
   Perhaps Alexanderson himself led an unexciting life. Perhaps he
had little time for excitement, producing as he did over 340 U.S.
patents, the last of which was issued to him a few years before his
death, at 97, in 1975.
   This thorough biography includes an index, extensive notes, a list
of Alexanderson's patents and about 20 photographs and
illustrations. For other reviews of it, see *OTB* August 1992, page 59
and August 1993, page 62.

All-American Five Radio: Understanding and Restoring Trans-formerless Radios of the 1940's, 50's and 60's
Richard McWhorter, Sonoran Publishing, LLC, 92 pages, 2003.
OTB, v44 n4, November 2003, p. 31-32.

What a neat idea for a book! McWhorter attempts to introduce the neophyte to radio receiver operation and repair by using a particular class of radios, in this case the ubiquitous All-American Five, exemplified here by the Zenith model 9X561/2. "DON'T PLUG IT IN" are, appropriately, his first words in chapter one. Good advice for the enthusiastic antique radio newbie.

McWhorter then gives the reader lightning-fast lessons in DC, AC, resistance, Ohm's law and capacitance before taking him methodically through the radio itself, beginning with the power supply, then moving backwards from the audio amplifier to the antenna. Schematic symbols are clearly shown and explained and the reader is often invited to compare the schematic with the corresponding equally clear pictorial drawings of each radio section.

It all works, but the book probably should have been longer or not tried to do so much. There just isn't enough room in the 45-or-so pages devoted to explanation of the radio stages, especially when one is writing to someone who may have come to the book with little or no knowledge of electronics.

The glossary should have been edited more rigorously or else omitted. It uses undefined technical terms such as *impedance, electromotive* and *coulomb* to define other terms. *Duplex-diode* is defined but *diode* is not. *Kiloampere* is defined (needlessly) but *kilohm* and *megohm* are not defined.

Some definitions simply make no sense, as "signal grid, the grid placed between the screen grid in a pentagrid vacuum tube," or "intermediate frequency . . . A frequency to which a signal wave is shifted location has an intermediate step in transmission or reception." This makes pretty rough going for the average inexperienced reader.

Altogether this book is a good attempt at explaining many complex concepts in very few words. Much can be learned from it and it will serve its purpose if it leaves the reader wanting more, but it must not be taken for a basic or introductory text.

Antique Radio News: A Bi-monthly Publication Dedicated to the Culture and Images of Vintage Radios
Number 1, English edition, May 1993.
Pittsburgh Oscillator, v8 n3, September 1993, p. 18.

This new magazine immediately takes its place in the long tradition of fine Italian color printing. The nine-page "Gallery" section contains photographs of early American consoles (RCA, AK, Zenith, Philco and Garod) and a 1909 German ship spark transmitter. It is difficult to imagine how these splendid photolithographs could be improved, but their images are, in fact, futher enhanced by being printed on "antique" paper, so that even though it is brand new, the magazine has the appearance of having moldered for many years in a museum library. Judging by the color photography alone this magazine is easily worth the subscription price.

But it contains more than pretty pictures. The Gallery section is followed by a Historical Figures section, with this issue featuring descriptions of world-wide reactions to the announcement of Marconi's death in July 1937. Here we find a description of the condolences sent from Berlin by Mr. A. Hitler. (Did he have any Atwater Kents?)

A preview of sections to be covered in the second issue includes "Gallery"–Thermo-ionic tubes; "Historical Figures"–Marconi's Death, A Look at the International Newspapers; "Top Radio"– Baird's Televisor; and "Talking About Radios"– Fleming's Valve by Domenico Ravalico. Evidently these broad section headings are to be included in each issue. The second issue is probably already in print, though we have not seen it.

Every antique radio editor wants to capture the "flavor" of radio history. One attempt to do this in *Antique Radio News* resulted in the section called simply "Story". Here the editors have chosen to reprint in installments Bawdry De Saucier's "The Beginning of Wireless Transmission", which originally appeared in the French magazine *L'Illustration* in 1923. This extensive article, translated into English, is to be serialized in future issues. Breaking points between the installments are quite literal, the one in this issue being in mid-sentence.

Yes, there is a "Buy-Sell-Swap" section, with advertisement space offered–apparently free–to "private individuals". It occurs to us that this might be the place for U.S. restorers of foreign sets to advertise for parts, information, schematics, etc.

We welcome this beautiful publication to the field of antique radio literature and wish it a long life.

The publication's address: Mosé Edizioni Via Bosco, 4 - 31010 MASER - TV - Italy.

Antique Radio Restoration
Bret Menassa, DVD Editions, 2001, 2004, 2005, 2006.

Volumes 1-2: *Pittsburgh Oscillator*, v19 n4, December 2004, p. 9.
Reprinted in *AWA Journal*, v46 n2, April 2005, pp. 56-57.

Volume 3: *AWA Journal*, v47 n2, April 2006, pp. 24-25. Reprinted
in *Pittsburgh Oscillator,* v21 n3, September 2006, p. 16.

Volume 4. *AWA Journal*, v48 n2, April 2007, pp. 37, 39.

Volumes 1 and 2: If you've been thinking about trying to restore
some of your old radios but hesitated because you "don't know
electricity," these two videos would be a good place for you to start.

Bret Menassa, looking very much like everybody's uncle in his
casual dress and with his gorgeous collection of radios in the
background, talks and walks the viewer through old radio repair in
a very laid-back informal style, assuming at the outset that the viewer
knows nothing about the subject. He begins with such rudimentary
lessons as knob, tube and chassis removal and cleaning methods.

Before the radio gets plugged in he shows how to change the filter
capacitors (they are assumed to be defective without testing) and
gives a very basic talk on tube operation theory. Only at that point is
the radio turned on. Much to Menassa's credit, he chose his
demonstration radio–a 1936 Philco Model 640–without knowing
beforehand what its operating condition was.

Those who already know the stuff that Menassa covers to this point
may still gain much from this video for he then spends considerable
time on cabinet repair and restoration. We are glad to see that he does
not strive for perfection–in fact all his cabinet restoration work
avoids refinishing, i.e., lacquer removal or stripping. The results he
obtains using only such products as oil soap, 000 steel wool, mineral
spirits, gel stains and "blush eraser" are stunning enough to make one
want to have his own 640.

One is tempted to recommend that the viewer spend some time
with a basic electronics text before viewing volume 2. It goes into

much greater depth in circuit analysis and troubleshooting methods. Menassa obviously is highly motivated in his desire to pass along his enthusiasm and knowledge to the viewer. It is very ambitious (some would say fool-hardy) to attempt to describe in just a few minutes such complex topics as capacitance, resistance, inductance, heterodyning, etc. in a way understandable to the neophyte. Engineers and electronics professionals will grimace at some of these explanations, particularly when they encounter such phrases as "soaks up the voltage." But of course engineers are not the videos' intended audience, and the motivated beginner can learn much from both videos.

Volume 2 ends with a detailed demonstration of a.m. radio alignment procedure using a signal generator. This is done by ear (no meter is used) and is an adequate method for the small table radios featured in the videos. Some detail should have been given about the connection between generator output and radio input.

These two videos represent an enormous investment of time and thought on the part of the narrator and producer. They are a viable alternative to the many available books and magazine articles covering the same material. Highly recommended to the wandering novice, and others as noted.

Volume 3: Okay, you've got your 1937 Emerson BA-199 in perfect working order, but–Holy Plaskon, Batman!–there's a one-inch chip missing from the cabinet. Your 1939 Belmont 636 is proudly featured on your mantel, but the brown Bakelite case is looking more than a little dull. At the flea market you found a sacred Fada "Bullet" (a steal at $400), but–*sacre bleu!*–there's a four-inch crack running across the top of the cabinet.

Get the picture? If you do, you need to see volume 3 of Bret Menassa's *Antique Radio Restoration*. Here you'll find detailed guidance on cleaning, restoring and repairing Bakelite, Beetle, Catalin, Plaskon, urea and modern plastics radio cabinets. (Okay, I admit it–I hadn't even heard of Plaskon until I saw this video). But the point is, anyone who has a radio with these or other no-nos in the cabinet who was previously resigned to living with the blemish now need not do so.

In fact, this video leaves the viewer with the idea that there is virtually no plastic cabinet insult that cannot be repaired. For example, halfway through the restoration of a 1930s Airline Bakelite cabinet, the piece was accidentally knocked off the workbench onto the floor (off camera). Bret chose not to record what his exact words were at the time, but he does show the pile of parts that resulted. It looks like nothing more than wastebasket fodder, but–you guessed it–the parts were again reassembled, reglued, refinished and repainted, with a final result that is difficult to believe.

This is non-wood radio cabinet restoration from A to Z from the man who knows how. Included are directions for molding Bakelite parts from epoxy resin and Bakelite dust. Volume 3 is done with the same knowledgeable care as the previous two.

Volume 4: In the first three volumes of this outstanding series Bret covered repair of radio innards and minor-to-major repair of radio cabinets made of wood and a variety of man-made materials. To his credit, Bret points out at the beginning of volume 4 that the first three volumes may be all you need to restore your radio cabinets. But if you are set on major cabinet repair, such as complete refinishing, then you should see volume 4.

Among other subjects, volume 4 includes repair and replacement of veneer, using store-bought veneer and "donor" veneer from junker cabinets. Bret demonstrates the ease with which veneer can be removed from cabinets. Reapplying veneer, especially to curved surfaces, takes somewhat more effort, but can be satisfactorily accomplished with care.

Bret also advises beginners to start by getting cheap junker cabinets to work on, cabinets that you can live without. As your experience grows you can tackle more valuable "keeper" cabinets.

In finishing and refinishing wood, Bret sings the praises of sanding sealers (tone finishes) and blush erasers, both of which are available in aerosol cans at your local home project mega-outlet.

The two table radios that volunteered for restoration on this disc are a Ward's Airline and a 1938 classic Zenith "glass rod" model 6-

D-317. At the end of the documentary one wonders, *Does Bret's glass rod Zenith work?* And then immediately thinks, *Who cares? Just* look *at it!*

(See also *Vintage Radio Alignment*, page186).

AWA Review
Volumes 1-5, CD Edition, Antique Wireless Association.
OTB, v43 n3, August 2002, p. 53.

The first five volumes of the *AWA Review* are out of print and unlikely to be reprinted in paper form. They get more scarce with each passing AWA convention.

Help has arrived. Even if your collection is missing only one of these volumes it makes sense to buy this CD to fill the gap. The CD appears to have been produced with the same care and at the same quality level as the CD version of the *OTB*. Requirements are a PC or Mac with Adobe Acrobat 4.0 or greater. The PC version of Acrobat Reader is included on the disc.

Though unfortunately the disc is not searchable at the single word level, it does include Ludwell Sibley's thorough index file, which can be searched by word. As a bonus, Lud's index covers all AWA publications through about the year 2000.

A big advantage of having the *Review* on CD is what librarians call file integrity. You're assured of no missing pages, no missing articles. At 741 pages total–something over two cents per page– this is bargain amateur history stuff.

Basic Radio Course
John T. Frye, 224 pages (1961 edition) , 1951, 1961, 1974.
Pittsburgh Oscillator, v17 n1, March 2002, p. 14. Reprinted in *OTB*,
v43 n4, November 2002, p. 46.

  So, you're new to the hobby and you want to know how old radios
work. Here is a good place to find out. After reading this book you
might want to move on to Frye's *Radio Receiver Repair*, reviewed
below.

  In relatively non-technical language–and with almost no math–Frye
explains how each stage of a tube-type radio works. Despite the
unfortunate error on page 16 of the 1961 revised edition which has
amperes equal to ohms divided by volts, the explanations are fairly
easy to understand for those willing to take the time to think about
them. Frye takes a somewhat unusual approach in that he starts at the
radio speaker and works backward to the antenna.

  If the author's name sounds familiar it may be because he was the
creator of "Carl and Jerry," the engaging radio boys' series that
appeared in *Popular Electronics* in the '50s and '60s. He also wrote
the "Mac's Service Shop" column for *Radio and Television News*.
Some of his hundreds of articles were translated into Spanish,
Portuguese, Japanese and Hindustani. He was born in 1910 and
began radio servicing in the late '20s.

Bibliography of Sir Oliver Lodge F. R. S.
Theodore Besterman, Oxford University Press, 219 pages, 1935.
AWA Journal, v51 n2, April 2010, pp. 58, 62.

It took Theodore Besterman, the great British bibliographer, 186 pages just to list the publications of Sir Oliver Lodge, one of the inventors of radio. Besterman is known by most librarians for producing his monumental *World Bibliography of Bibliographies*. To say that he was thorough is a gross understatement. How thorough? This will give an indication: he lived in a former house of Voltaire's while he worked for years on his extensive Voltaire bibliographies.[1] He produced, in 107 volumes, an edition of Voltaire's correspondence numbering about 20,000 letters.[2]

For some reason unknown to me, Besterman chose to number Lodge's publications using Roman numerals. So, for example, we have Lodge's publication number DCCCCXCIX (this would be number 999 for us non-Romans). That was Lodge's "The Problem of Television," which appeared in the March 1928 issue of *Popular Wireless*. (I'd love to know what Lodge thought television's problems were in 1928, but have not yet been successful in finding this periodical.)

Besterman published his Lodge bibliography in 1935, five years before Lodge's death. The last publication listed for Lodge was number MCLIV (1,154) for "The Progress of Electrical Science" in the January 5, 1935 issue of *Nature*. Lodge published more than 200 articles in this prestigious science periodical. Most of his articles on psychical research, however, appeared in *The Journal of the Society for Psychical Research* or in the Society's *Proceedings*. (Besterman himself was greatly interested in psychical research).

Lodge contributed a foreword to Besterman's bibliography. What would you choose to write about in the foreword of the book that listed your 1,154 publications? Tellingly, Lodge chose to write about one of his greatest *non*-publications, i.e., his demonstration of radio before the British Association on August 14, 1894. After describing his demonstration in the foreword, he wrote:

This was admittedly a very crude beginning to what has become a large method of signalling. Two years later, in 1896, it was brought to public notice by Marconi and Mr. Preece working hand in hand with Post Office facilities. But, so far as I know, my experiment was the starting-point of Sir Henry Jackson's signalling between ships by means of Hertzian waves, and of Professor Popov's apparatus in the Russian Navy for the same purpose . . . . Signor Marconi made a step of real novelty and epoch-making importance when he heard signals from Poldhu away in Newfoundland, and found that the waves' progress was not interfered with by the curvature of the earth, but that they could travel even to the Antipodes.

Lodge's contributions to radio were largely overlooked by the public after Marconi came on the scene, and Lodge wanted here to set the record straight, though ever remaining the Victorian gentleman.

1. *http://en.wikipedia.org/wiki/Theodore_Besterman*
2. *http://www.nybooks.com/articles/11460*

Boy Genius and the Mogul: The Untold Story of Television
Daniel Stashower, Random House, 277 pages, 2002.
OTB, v43 n3, August 2002, p. 55.

Historians will be disappointed with this book. It lacks footnotes and contains only a two-page bibliography that lists mostly secondary sources. A short list of periodicals shows titles only. The book's subtitle is less than accurate and was probably added for sales appeal.

But Stashower is a good writer who tells a good story. Philo T. Farnsworth certainly counted as a boy genius–he independently conceived of electronic television as a high school boy of 14. And Sarnoff certainly counted as a mogul–he forged RCA and then used it to forge the mid-century telecommunications world. Stashower's story of the professional struggles between these two men includes thumbnail sketches of Alexanderson, Armstrong, John Baird, de Forest, Elma Farnsworth, Gernsback, Charles Jenkins and Zworykin.

Competition, we are told, is good for us. Seldom is the caveat *up to a point!* added. The competition between Farnsworth and Sarnoff almost certainly elicited ruthlessness in Sarnoff–"I don't get ulcers; I give them"–and certainly had a deleterious effect on Farnsworth's health and fortune.

Sarnoff also liked to say, "The Radio Corporation does not pay royalties, we collect them." In fact, Farnsworth was able to force RCA to pay royalties to him for his invention of the image dissector. But although Farnsworth won the patent battle he lost the television war.

The book might be summed up by the Charles F. Jenkins's quote which it contains: "It's the old story over again. The inventor gets the experience and the capitalist gets the invention."

Another view of the Farnsworth and Sarnoff fight is Evan Schwartz's *The Last Lone Inventor: A Tale of Genius, Deceit and the Birth of Television*, HarperCollins Publications, 2002, 322 pages.

Candid Autocamera Biography
George H. Clark, Reprinted by TCA Inc., 30 pages, 2002.
OTB, v44 n1, February 2003, p. 33.

   This little publication brings to the surface the story of a radio pioneer many of us would never have heard of otherwise. Clark (1881-1956) was a character, all right. Soon after John Stone hired him as an assistant in the early 1900s, Stone realized that Clark's math talents were somewhat deficient and offered to pay his way through a math course at MIT. But after a few months Clark dropped out, realizing he "hadn't the faintest idea of what it was all about."
   Later Clark worked for years in a variety of odd jobs with RCA. During one of these he was sent to Venezuela (he knew no Spanish) to install a series of transmitters. It was while he was there that RCA chose to "give him the sack." Clark solved that problem by simply returning to the U.S. and continuing to work for RCA. For its part, RCA simply continued to pay him!
   The hour-long CD interview included leaves the listener wanting more, particularly more about all the radio pioneers that Clark knew personally–Armstrong, Stone, de Forest, Miessner, Kolster, Weagant, Sarnoff, etc.
   But surely Clark's biggest contribution to radio history resulted from the fact that he was a saver. Generally, if it was written or printed and about radio, if Clark got his hands on it, he saved it. This included entire libraries of primary material from RCA, Hammond, Jenkins, Farnsworth and others. Much of the final collection, too vast even to outline here, eventually wound up at the Archives Center of the Smithsonian National Museum of American History. You can see a detailed contents listing of that material at:
   www.si.edu/lemelson/dig/radioana/#bio.

Carl & Jerry: Their Complete Adventures from Popular Electronics
John T. Frye. Edited by Jeff Duntemann. Copperwood Press, five
  volumes, about 190 pages per volume, 2007.
AWA Journal, v48 n4, October 2007, p. 58.

Yeah, you're gonna like these, all right. Even in the unlikely event that you've never read any of the adventures of Carl and Jerry, you'll still enjoy reading these stories of two teenage boys, each gaga over electronics, learning the ins and outs of the hobby by the seat of their pants in small town America at mid-20th century. These 119 stories originally appeared in *Popular Electronics* between 1954 and 1964.

Seen from the sophisticated electronic perspective of the 21st century, these short stories may strike the reader as somewhat pedestrian. And some of the stories stretch the limits for short fiction. "Benefits of Amateur Radio," for example, is essentially a list of benefits very thinly disguised as a short story.

Frye knew he was onto a good thing, and repeats some techniques and descriptions. How many ways can he describe Jerry as fat, without actually using the word? And how many times must we be told that Carl wears horn-rim glasses? So the writing is formulaic, of course, but who cares? It's fun to see what entertained us and taught us half a century ago.

With the editing of this five-volume series now under Duntemann's belt, it should be a relatively easy matter for him to collate and print Frye's "Mac's Service Shop" series next. We're waiting.

A word of warning–you've heard about the kid set loose in a candy shop? The first two pieces were great, the next two were merely okay. Surprisingly the fifth wasn't too good, and after that . . . . Carl and Jerry comin' atcha once a month made us look forward to each episode, but it would not be wise to read these books straight through at once. A workable compromise might be to read one story a day for each of the next 119 days.

Charles Herrold, Inventor of Radio Broadcasting
Gordon Greb and Mike Adams, McFarland & Company, Inc., 247
  pages, 2003.
OTB, v44 n4, November 2003, p. 31.

Mike Adams's companion film *Broadcasting's Forgotten Father*
appeared in 1994. Now he has teamed with broadcast educator
Gordon Greb to produce what should long remain the definitive work
on Herrold.

The book actually has two subjects. It is a thorough and detailed
biography and it is an extended discussion of the endless–and
endlessly interesting–question of who was first.

Born in 1875, Herrold used a Rhumkorff coil and Branley tube in
1895 to broadcast a telegraph signal for a mile after reading a news
item describing Marconi's experiments. By 1909 Herrold was
broadcasting on a regular basis using a patented arc transmitter of his
own design. He notarized a description of his broadcasting work and
publicized it in a radio catalog in 1910. He spent much of his later
years attempting to gain recognition for his early work. But lacking
the charisma and self-promoting ability of another broadcasting
pioneer, Lee de Forest, Herrold slipped out of the public eye and died
unknown in 1948.

Simultaneous or independent invention or discovery is nothing
new. Newton and Leibniz invented calculus independently of each
other in the 17th century (and argued over who was first for much of
the rest of their lives). So too in the early 20th century several men
invented radio broadcasting in ignorance of each other's work.
Everyone's familiar with KDKA's pioneering work in 1920, but few
know of Charles Herrold's much earlier daily broadcasting in San
Jose, California in 1912 and later. So, were KDKA and Frank
Conrad the KCBS (descended from Herrold's station) and Charles
Herrold of Pittsburgh or was it the other way around?

In what probably won't be the final analysis of the subject, the
authors allow that KDKA is the oldest broadcasting station but that
KCBS was the first. Their case for Herrold's primacy in broadcasting
is cogently presented and difficult to refute.

Chemical History of a Candle
Michael Faraday, 1861, reprinted by Cherokee Publishing, 192
    pages,1978.
OTB, v43 n4, November 2002, p. 45.

This classic Faraday essay introduced young people to science through his ingenious examination of the actions of a simple burning candle. It has nothing to do with radio but will be of interest to anyone having a technical bent. This is how all science should be written, though for many practical reasons it cannot be.

Listen to Faraday in a lecture-demonstration to youngsters 100 years before the invention of OSHA:

> *Here is a little gunpowder. You know that gunpowder burns with flame—we may fairly call it flame. It contains carbon and other materials, which all together cause it to burn with a flame. And here is some pulverised iron, or iron-filings. Now, I purpose burning these two things together. (Before I go into these experiments, let me hope that none of you, by trying to repeat them, for fun's sake, will do any harm. These things may all be very properly used if you take care; but without that, much mischief* will be done).*

* An example of such "mischief" was to be graphically and hilariously depicted years later in Chopin's biopic *Impromptu*.

CK722 Classic Germanium Transistor Website
http://www.geocities.com/ck722transistor/
OTB, v43 n4, November 2002, pp. 46-47.

Some people are passionate about women. Others are passionate about men. Still others (like us) are passionate about radios. Then we have those who are passionate about *parts* of radios. Enter Jack Ward, who has spent many hours researching the history of transistors–one in particular, the famous CK722 produced in the '50s and '60s by Raytheon.

Ward has just finished a 120-page book, *The Story of the CK722*. Visit the site above for details and many CK722 links, including contributions by those who were in on the design of the CK722. The book is available with two CD's, which provide supplemental material.

Ward's book includes a modern-day solid-state "Carl & Jerry" adventure, written by himself. I won't give the ending away, but, as you might expect, the CK722 plays a pivotal role.

I'd like to see all of John Frye's original Carl & Jerry adventures and Mac's Service Shop episodes reprinted. Is anyone passionate about that idea?

(See *Carl & Jerry*, page 17).

Classic Cones: Pictorial Reference and Value Guide for 1920s Radio Cone Speakers
Buford and Jane Chidester, Sonoran Publishing, LLC, 122 pages, 2001.
OTB, v43 n2, May 2002, pp. 51-52.

As the hobby ages it is refined. Refinement frequently means specialization. Chidesters' book is a good example.

First there were only headphones, then came horn speakers. These provided needed volume but not much increase in frequency response. The next improvement, from about 1924 through 1929, were cone speakers. Instead of the small diaphragms of headphones and horn speakers, cone speakers used vibrating surfaces of from four to 36 inches and thus provided a wider frequency range than headphones could hope for.

Suggested current prices are offered for most of the 250-or-so cone speakers shown; some are listed only as rare or very rare. Some reproductions of original advertisements are included. As indicated by the subtitle, there is little text to this book, so don't look here for technical information or history.

Photographs are in full color, and this helps show that manufacturers' attempts to make these speakers attractive, as well as functional, fell along a wide scale. Hence, some of these speakers will be perceived as beautiful, others as ugly, still others as so ugly they're beautiful–like some of us think of Atwater Kent Model 40 radios.

This book is typical of the high quality we have come to expect from the Sonoran Company. Visually, it neatly takes up where the many horn speaker contributions by Floyd A. Paul leave off.

Collector's Guide to Antique Radio
John Slusser *et al*, Collector Books, 285 pages, 2001, 5th edition.
OTB, v42 n4, November 2001, p. 29.

Buy this book even if only to get a copy of the reproduction of Jim Daly's artwork on the front cover. Here slouches dreamy-eyed Billy America, listening to the family's brand-new 1932 Philco 71B roundtop. Billy needs a haircut, his shirt and bib overalls must soon see a washing machine, and at least one sole of his shoes needs to be covered again with a piece of cardboard, but all these matters can wait. Right now the score is tied and its getting pretty late in the ninth.

More nostalgia awaits us inside, incited by the beautiful color photographs of the We-had-one-just-like-that-when-I-was-a-kid! variety. The book follows the now-familiar style of the first four editions by Marty and Sue Bunis. Each entry contains a two- or three-line description of the radio, ending with a suggested current retail price on the flea market or garage sale scene.

Slusser provides a schema for his determination of suggested prices in the front matter of the book. Suggested prices are sure to cause some disagreement and conversation, and that is one of the values of the book. A Philco 71 is listed in the first (1991) edition of this series for $175, in the 1992 edition for $250, 1995 for $250, 1997 for $200-250, and current edition for $230-280. Are these prices fair? Are they accurate? Are they even realistic? Each buyer or seller will have to determine that for themselves.

Collector's Vacuum Tube Handbook: Volume I: The Non-RMA Numbered Receiving Tubes
Robert T. Millard, Sonoran Publishing, LLC, 196 pages, 2001.
OTB, v42 n3, August 2001, p. 30.

One is immediately struck by the clear, logical layout of this book. For most tubes covered there is a photograph of the tube and of a typical tube carton, a base diagram, a brief historical paragraph (which often indicates the rarity of the tube), and a table of maximum ratings and another of typical operation figures. Most data have been taken from manufacturers' sheets or tube manuals, but when no data could be found in some cases the author generated it using an AVO Mark IV dynamic mutual conductance tube tester.

Presumably future volumes in this series will cover tubes with RMA (Radio Manufacturers Association) numbers. Lack of standardization in tube numbering can obviously lead to confusion or worse. The author points out that although most 01A's are interchangeable, an RCA UX 222, Sylvania UY 222 and an Arcturus AC 22 are radically different tubes. We are reminded here of the old adage about standards–the nice thing about them is that there are so many to choose from.

Even if you don't need the technical statistics provided, you may still want a copy for the interesting tidbits found in the historical descriptions. This may include tube function, manufacturer's name, date of introduction, reason for new type, complementary and competitive types, envelope type, time period during which the tube was popular, reason for demise, substitutions for type, and whether the tube had any unusual features, such as double filaments, use of screen plates, etc. The book includes a short but appropriate bibliography and an index by tube number.

Most tube collectors and tube historians will want to own a copy of this first-rate reference book. *Note: published book not seen by reviewer.*

Communications: An International History of the Formative Years
Russell W. Burns, Institution of Electrical Engineers, 639 pages,
  2004.
AWA Journal, v46 n2, April 2005, p. 56.

Anyone seeking an in-depth introduction to the subject of communication by electromagnetic waves would do well to consult this volume. It is hard to believe that just one man wrote this thick book. Burns achieves his goal by relying heavily on secondary sources, many of which are his own. (He is the author of four other books in the IEE series). Though some of the information here can be found in other sources, Burns provides a valuable service in bringing it all together in one comprehensive compilation and interpreting it.

He begins his big history with a chapter on communications among the ancients. There follows a chapter on the instrument which led (he coyly but correctly claims) to the "first practical, long range, communication system using electromagnetic waves"–the telescope.

Three examples will show the depth of detail in the book. We are told on page 146 that in his attempts to finance a telegraph cable between Europe and America in mid-19th century Cyrus Field crossed the Atlantic Ocean 64 times, getting seasick each time.

One probably wouldn't expect to find a photograph here showing the primitive but effective "method of releasing a carrier pigeon from a porthole in a [World War I military] tank", but there it is on page 409. (We can assume this photograph is no longer classified).

A further example is the "Summary of the proposals made during the period 1911-1930 for electronic television cameras using cathode-ray tubes." This seven-page table lists no fewer than 22 such proposals from five countries, with descriptive details for each.

Burns interprets "formative years" in his subtitle broadly. So we have a chapter on high definition television (meaning CRT television) and a chapter on radar. AWAers may be most interested in the chapters "The birth of sound broadcasting," "Some important developments in the 1920s," and "Rise and fall of low definition television."

Our only complaint is about the reproduction of some photographs from other publications. The details in some of these are so tiny as to border on the unreadable.

This book is something like a great accomplishment; it makes informative reading.

Communications Miracle: *the Telecommunication Pioneers from Morse to the Information Superhighway.*
John Bray, Plenum Publishing Corp., 379 pages, 1995.
Pittsburgh Oscillator, v14 n2, June 1999, p. 13. Reprinted in *OTB*, v43 n4, November 2002, p. 47.

Reading this book was a genuine pleasure. It is written in language that is always clear and generally concise, it is very well-organized, and the author, a British telecommunications engineer, witnessed firsthand many of the events and inventions he writes about.

In the introduction to the book Bray states: "The present-day world telecommunication network is the most complex, extensive and costly of mankind's technological creations and, it could well be claimed, the most useful." The pages that follow document this claim in detail, beginning with the creation of the mathematical foundations of radio by Volta, Ampere, Ohm, Faraday, etc., and continuing through a chapter on telecommunications and the future.

We get chapters on telegraphy, the telephone, radio, television and microwave transmission, but my eye was caught by the fascinating details about such major engineering endeavors as the first successful trans-atlantic telegraph cable (completed before the American Civil War), the first trans-atlantic telephone cable (1956) and the first trans-atlantic optical fiber cable (1988).

The first telegraph cable provided no signal amplification, of course, and the distributed capacitance of the 2200-mile cable severely limited the speed at which messages could be sent. Nevertheless messages were exchanged between Queen Victoria and U.S. President Buchanan on August 14, 1858.

A century later the first telephone messages crossed the Atlantic by cable (actually two cables, one for each direction of transmission). Even though the transistor had been invented in the late '40s, these cables used tube amplification, with 146 repeaters (amplifiers) each using tubes designed to operate for 20 years without attention. Bray refers to the 1956 system as "the most outstanding engineering achievement in telecommunications during the first half of the 20th century." The cable provided 35 telephone circuits between London and North America. A mere 32 years later the world's first optical fiber transatlantic cable provided 40,000 circuits.

Trans-atlantic telephone cables with transistorized repeaters were not installed until the 1960s, and reliability standards for the transistors used were ultra-high. Bray states: ". . . typically there must be not more than one failure in 2000 transistors over a period of 25 years!" Transistors were tested for reliability using "accelerated aging tests" involving operation of the transistors at above-normal temperatures. (I have always objected to the name of this test, since all material always ages at exactly the same rate–24 hours a day–no matter what stresses it is subjected to, except maybe in black holes. Perhaps a better name would be accelerated stress test).

Bray finishes his book with "And Part of Which I Was," an essay written in the genre I call technobiography, i.e., biography as revealed through the technological innovations or experiences of an individual's life. Technobiography has been practiced simultaneously with the history of telecommunication, and includes the writings of de Forest, Fessenden, Fleming, Lodge and many others. These writers give us personal descriptions of radio history as it happened, made all the more interesting by being direct eyewitness and earwitness accounts.

Complete Jackson 648 Tube Tester Manual
John Cross, Vacuum Tubes, Inc., 92 pages, 2004.
OTB, v45 n3, July 2004, p. 13.

In the beginning there were books about radio, then there were books about radios. Soon there appeared books about test equipment, and now there are books about particular models of particular types of test equipment. Jackson tube tester owners take note: Cross's compilation of Jackson facts brings the relevant *648* documentation all together in one convenient place. Cross includes history and development, operating instructions, setups for antique tubes and radio restorers, and service information and schematics.

The main feature is a 55-page reproduction of the complete roll chart for the *648*. Because the book is spiral bound and lies flat when open, the roll chart is very convenient to use. In Army basic training I was keen to avoid the dreaded "M14 thumb." Cross's inclusion of *648* roll chart settings allows users to avoid "roll-chart thumb." This feature alone makes the book worth purchasing. Recommended to all *648* users.

Cooke and Wheatstone and the Invention of the Electric Telegraph
Geoffrey Hubbard, Routledge & Kegan Paul, 158 pages, 1965.
AWA Journal, v52 n1, January 2011.

Appropriately, this book was written by someone who began his career in science and then transferred to administration. I say appropriately for it was Hubbard's main task in the book to explicate the complicated, sometimes antagonistic and bitter, relationship between the essential scientist Charles Wheatstone and the essential practical administrator William Fothergill Cooke. So, just like in telegraphy, there were bound to be some sparks in the Wheatstone and Cooke endeavor. To say the least, they made an unlikely partnership, or to quote Hubbard:

> *It seems then, that each partner was essential to the other; Wheatstone for his scientific ability and Cooke for his driving force. Their common tragedy was, that though each was likeable in his own way and had many devoted friends, they were mutually antipathetic; the collaboration which enriched their era poisoned their lives.*

Hubbard contends that the technological and social situation in the England of the 1830s was so ripe for the invention of the telegraph that if Cooke and Wheatstone hadn't invented it, someone else soon would have. He names Francis Ronalds, Professor Steinheil and Baron Schilling as inventors who were rapidly closing in on the telegraph before Wheatstone and Cooke. But Schilling abruptly quit his telegraph researches, even when he was within grasp of solving his last technological problem. The reason? Schilling had been, to use the quaint 19th century British phrase, "surprised by death."

Cooke's and Wheatstone's first telegraph patent was issued in England in 1837 (signed by no less than King William the Fourth a few days before he died), while Samuel Morse's appeared in the United States just three years later. So Cooke and Wheatstone were the first, but within a few years their complicated five-needle telegraph was supplanted in England by the simpler recording telegraph of Morse.

One can imagine that the aloof Wheatstone could be frustrating to work with. At one point in the development of the telegraph, he took time off to invent the stereoscope, known in today's form as the child's toy called the View-Master™. Hubbard also credits Wheatstone with invention of the dynamo, the concertina, and one of the three viable forms of the typewriter.

Hubbard expresses gratitude in his book that the telegraph was invented before the telephone. Thus Cooke explained his invention in lengthy letters to his mother (rather than just speed dialing her cell). Many of those letters still exist, and Hubbard quotes extensively from them in the book. He also used them to advantage when he ventured into explaining the psychology of the Cooke/Wheatstone relationship.

Geoffrey Hubbard was an accomplished writer, astute and careful; it is a joy to see him at work. He quotes from another Cooke letter back home to Mom: "Did I tell you that Mr. [Isambard Kingdom] Brunel took me in his carriage to Maidstone on Thursday week, to view his railroad and works?" Hubbard succinctly corrects Cooke: "But it must have been to Maidenhead, not Maidstone, that they went" (p. 71). And he does not shy away from criticizing other primary sources in the book, as when he clearly explains a new patent issued to Cooke and then states that with it Cooke had invented a telegraph that was actually less useful than the one he already had.

Despite all of Cooke's and Wheatstone's disagreements and antipathies, their first practical telegraph was finally installed in 1838 on the Great Western Railroad. Some locals and VIPs were impressed, but the public at large knew little of it or cared little about it. However, interest was greatly boosted when the telegraph was used to catch a murderer, John Tawell, in 1845. The book gives the details in a six-page chapter. History often repeats itself. Sixty-five years later, the public's interest in radio was greatly increased by the famous use of radio telegraphy to catch another British murderer, Dr. Hawley Crippen.[1]

It is known that great people, just like us ordinary folks, can be small minded. Europe gave many medals and diplomas to Wheatstone in recognition of his accomplishments. But he refused to accept the Albert Medal offered to him by the Royal Society of Arts. Hubbard explains: "The Society made the mistake of offering the same honour to Cooke . . ."

Hubbard ends his book on a sad note:

> *A hundred years have passed. The telegraph is no more. Nobody remembers William Fothergill Cooke, and Charles Wheatstone is remembered, if at all, for the Bridge which he did not invent.*[2]

Sic transit gloria mundi. Still, faint–though clear–echoes of Wheatstone's and Cooke's long-ago struggles with hardware and with each other could be heard as recently as the August 2010 World Convention in West Henrietta, New York as AWA members demonstrated their fists in the hallway of the RIT Inn for any and all who cared to listen.

Telegraph historians and enthusiasts–and those who like good biography–should not overlook this book.

1. See *http://www.titanic-whitestarships.com/The%20Dr.%20 Crippen%20Story.htm*
2. See *http://en.wikipedia.org/wiki/Wheatstone_bridge*

Crosley: Two Brothers and a Business Empire That Transformed the Nation
Rusty McClure with David Stern and Michael A. Banks, Clerisy
  Press, 504 pages, 2006.
AWA Journal, v49 n1, January 2008, p. 29.

Despite the authors' penchant for–indeed, their insistence upon–
literary fillips, this book is definitely a "good read", to use current
popular review jargon. In fact, the book has risen to the status of a
New York Times bestseller, and deservedly so. The fast-paced style
of the writing was, perhaps, meant to parallel the fast-paced style of
the Crosley brothers' lives.

The authors dedicate their book to "enabling freedom and
entrepreneurial courage, served by those whose ability and
conviction make achievement possible . . . ." This is entirely
appropriate for a book describing the careers of Powel and Lewis
Crosley. Their styles, abilities and motivations, though distinctly
different, complemented each other; together the two-man team
became a largely-unstoppable success story.

Powel, who failed to finish college, was a visionary idea man who
couldn't wait to finish one project before rushing off to the next.
Lewis, with an engineering degree, provided ballast by taking care of
the many practical details of Crosley business. He usually was home
by five to be with his family for dinner. Together the two built the
Crosley empire. The book mentions only some of the estates, yachts,
properties, factories, cars, planes, islands, etc. that Powel came to
own after his phenomenal success. He also bought the Cincinnati
Reds baseball club, vowing to keep it in Cincinnati as long as he
lived.

Most readers will associate the Crosley name with radios, but that
wasn't Powel's first love. His career began, and ended, with
designing automobiles. In between those two mini-careers were
radios and WLW. Crosley radio and WLW complemented each other
as did Powel and Lewis; like the two brothers, both became wildly
successful. Never one to do anything small when it could be done
large (the Crosley car and Pup were exceptions), Powel soon took

WLW's power to an unheard-of "experimental" 500,000 watts. (Today it is back to a more human 50,000). Powel believed–correctly–that the more power WLW had, the more Crosley radios would be sold.

The authors show the Crosleys' success by listing materials used by their radio manufacturing facilities in *one week*: ". . . three million screws, five million nuts, six tons of bus wire, sixty thousand binding posts, half a million square inches of Formica, fifteen thousand sockets, six thousand audio transformers and one railroad car full of cardboard cartons." By 1922 Crosley was the largest radio manufacturer in the world, and Powel became known as "the Henry Ford of Radio." Another of the Crosley brothers' strategies had paid off big time: "build for the masses, not the classes."

Diversity was another Crosley manufacturing strategy. Most readers will be familiar with the famous Crosley "shelvador" refrigerators, but not all will know of the company's all-time most profitable device, the Icyball–(you'll have to read the book).

This book appeals to car and airplane enthusiasts, students of 20th century American culture, radio nuts, entrepreneurs, baseball fans and the many who just like good biography, no matter whom the subject.

David Sarnoff Research Center: RCA Labs to Sarnoff Corporation
Alexander Magoun, Arcadia Publishing, 128 pages, 2003.
OTB, v45 n2, May 2004, pp. 26, 31.

At first glance this appears to be a thicker-than-usual RCA promotional pamphlet, but closer inspection reveals it's much more than that. This book affirms the wisdom of a well-known saying about the value of a picture. It amounts to a brief pictorial history of RCA Laboratories (now the Sarnoff Corporation) from its beginnings in 1941 to the present. The pictures are made even more worthy by the inclusion of insightful captions. These very often identify the persons pictured, so the book provides portraits of many lesser-known behind-the-scenes radio technologists seldom included in popular radio histories. (Too bad that a large proportion of the photos of women are unidentified).

We have, for examples, posed photos of James Harbord (RCA chairman), Browder J. Thompson (high-frequency tubes), James Hillier (electron microscope), George Brown (rf heating), Albert Rose (image orthicon), Harry Olson (acoustics), Les Floury (infrared tube), Richard Webb (electronic color TV camera), Jan Rajchman (magnetic core memory), Elmer Engstrom (Center director), Charles Young (fax technology) and many others. Of course Zworykin makes many appearances, including taking a bulldozer ride at the Research Center with Irving Wolff at the "steering wheel."

Thanks to Magoun for getting these photos all in one place and identifying so many of the RCA research pioneers.

The Diary of a Pilot: China–Burma–India, 1943-1945
Arch Doty, Jr., W7ACD, *booklocker.com, amazon.com*, 143
   pages, 2007.
AWA Journal, v49 n1, January 2008, pp. 29-30.

During the second World War the Japanese occupied large areas of eastern China, blocking land and sea routes to the interior. The only way to get military supplies to unoccupied China was to fly them from northeast India across the Himalayan mountains, known as "The Hump," into central China. First Lieutenant Doty made this trip many times between 1943 and 1945, carrying a variety of cargo in a variety of military planes. Now, at 87 years of age, Doty has decided it's time to publish the diary of his war pilot days, together with some contemporary comments.

At the end of his diary he includes a table listing many of his Army Aviation friends. A chilling feature of the table is a column which records the dates on which his friends were killed during the war–yet another confirmation of General Sherman's famous three-word dictum about war.

Between the horrors of war and the hazards of flying over mountains that were miles high, Doty occasionally found himself with time on his hands. He wanted to hear news about his homeland but from Jorhat, India his five-tube radio was unable to receive U.S. broadcasts. So he did what anyone else in his situation would do. He consulted his copy of the *ARRL Handbook* and began designing radio antennas. He scrounged 520 feet of wire and, using 50-foot high bamboo poles, constructed a V-shaped antenna aimed at San Francisco. The result was "we had KWIX so loud that we couldn't talk." KWIX was an early Voice of America station located in San Francisco during the War.

Doty must have studied that *ARRL Handbook* pretty thoroughly. His post-war career included, among many other activities, obtaining several U.S. patents, one of which is for an improved vertical antenna. You can see the patent by typing in 4658266 at:

www.google.com/patents.

Distant Vision: Romance and Discovery on an Invisible Frontier
Elma G. Farnsworth, PemberlyKent, 333 pages, 1990.
Pittsburgh Oscillator, v6 n2, June 1991, p14.

Some people become inventors accidentally while they are trying to do something else. Others work most of a lifetime before they achieve success. Then there was Philo T. Farnsworth, who, as a 15-year-old farm boy, already knew what he wanted to invent and how to go about it.

Farnsworth wanted to invent electronic television, and while in high school he made a rough sketch of part of his proposed invention for his chemistry teacher. Six years later on September 7, 1927 in San Francisco, Farnsworth, at the age of 21, demonstrated his working all-electronic television, using the "image dissector" tube he had designed and built.

In addition to his image dissector, Farnsworth invented and patented circuits for TV scanning, synchronizing, magnetic focusing, vertical deflection, maintaining constant black level, and generating high voltage from the horizontal scanning voltage. Of his 100+ patents–a list of which is included in the book–Farnsworth considered his two nuclear fusion patents to be most important, though they have thus far produced no practical devices.

Farnsworth seemed forever caught between his desire to do pure research in electronics and his desire to bring his inventions into the marketplace. His story parallels the stories of many electronics inventors in the early part of the century who seemed to have spent as much energy in patent litigation as they did in inventing. Years after he drew it, Farnsworth used the sketch he made in high school to help win his priority patent case against RCA and Vladimir Zworykin.

Distant Vision is the story of Farnsworth's life as told from the personal viewpoint of his wife Elma. Other views of his work are available in George Everson's *The Story of Television–The Life of Philo T. Farnsworth* (Norton, 1949) and in Albert Abramson's *The History of Television, 1880-1941* (McFarland, 1987).

E. H. Scott and John Meck Story
Author, publisher and date unknown, 37 pages.
AWA Journal, v46 n4, October 2005, p. 59.

The subject is the Scott Radio Laboratories in Chicago and the John Meck Industries in Plymouth, Indiana from 1949 to 1956. The pamphlet is subtitled "A Pictorial Look" and consists of photographs of the principle players, manufacturing plants and products. A list of Scott company directors and officers is included.

Our favorite picture is of a Scott advertisement showing "noted opera star" Lauritz Melchior relaxing in his backyard swimming pool while his smiling wife and a huge Scott radio-TV console just happen to be nearby on the pool's deck.

Early Days of Radio Broadcasting
George H. Douglas, McFarland, 248 pages, 1987.
Pittsburgh Oscillator, v2 n4, December 1987, p. 4. Reprinted in
OTB, v43 n4, November 2002, pp. 47-48.

George H. Douglas, born in 1934, did not live during most of the period being written about in this book; hence the flavor of first-hand experience is missing here. But this informal, sometimes anecdotal, history is well-written and will be interesting to anyone who wants to know more about the early days of broadcasting.

The book opens, appropriately enough, with a chapter titled "KDKA." This includes portraits of Marconi, Alexanderson, de Forest, the 16-year-old Sarnoff, and Frank Conrad at his work bench. Two other chapters complete the early general history, and these are followed by chapters on specific aspects of the subject: the radio announcer, radio advertising, regulation, news, sportscasting, networks, educational stations, classical and popular music, and "Amos 'n' Andy"–each chapter being a well-researched essay on the topic. This approach enables the reader interested in a particular aspect of radio history to quickly locate relevant material in the book, but makes the book as a whole difficult to review. We will look at only two chapters here.

In "The Wavelength Wars," the chapter on radio regulation, Douglas relates the heavy-handed but necessary measures taken by Herbert Hoover, as Secretary of Commerce in Harding's administration, to impose some order onto the chaos of broadcasting in the mid '20s. His efforts were not altogether appreciated. As proof, Douglas quotes the following excerpt from a letter Hoover received from female evangelist Aimee Semple McPherson:

> *Please order your minions of Satan to leave my station alone. You cannot expect the almighty to abide by your wave length nonsense. When I offer my prayers to Him I must fit into His wave reception. Open this station at once.*

If you're too young to actually remember the "Amos 'n' Andy" show, you'll want to read that chapter in Douglas's book. Gosden and Correll created the first serialized sitcom on radio, "Sam 'n' Henry," and later, "Amos 'n' Andy," radio's first syndicated program. Douglas details the phenomenal success of this program in the early '30s, during which it is believed that a third of the population of the country tuned this program in.

> *Drugstore proprietors refused to wait on customers while the program was in progress. Motion picture theatres installed radio loudspeakers in their lobbies, and sometimes stopped whatever picture was being shown so that their audiences could listen to the fifteen-minute show.*

Douglas includes a robust 19-page bibliography in his book. The material in it is divided into bibliographic sources, books and articles, unpublished material, government publications, periodicals, and broadcast archives; most publications listed deal more or less with the 1920s and 1930s era.

Early Development of Radio in Canada 1901-1930
Robert P. Murray, Sonoran Publishing, LLC, 154 pages, 2005.
AWA Journal, v46 n4, October 2005, p. 59.

The book reproduces about a dozen previously-published historical essays written by Murray and others.

It is immediately tempting to compare this book with Lloyd Swackhammer's *Radios of Canada*. Swackhammer's book covered radio manufacturers in alphabetical order. Murray's book emphasizes developmental trends and people–some, like Fessenden and Marconi, well known; others not so well known. There are also chapters on kit radios and radio patents.

Thanks to Roger Hart's and Robert Murray's essay we can throw XWA, Montreal into the mix of stations vying for the title of first broadcaster. The authors provide tantalizingly few details, but state: ". . . it does appear that XWA was the first to broadcast regular, scheduled programs" (p. 25). (Do Hart and Murray know of Greb's and Adams's *Charles Herrold, Inventor of Radio Broadcasting*?)

The text of these essays tells only half the story, for the book also contains about 200 clearly identified black-and-white photographs, with additional line art, schematics, charts and ad reproductions. The index (rare in a book of collected essays) appears to bring together references to most equipment, model numbers and people.

Lastly there is a little gem, unexpected in a book covering 1901 to 1930–an essay on the trials and tribulations of a 1940s Canadian radio manufacturer, Hank Thorkelsson of Thorcraft Radio. The essay is particularly interesting in its contrast to the subject of the preceding chapter–the Canadian activities of radio giant RCA.

Early FM Radio: Incremental Technology in Twentieth-Century America
Gary L. Frost, The Johns Hopkins University Press, 191 pages,
  2010.
AWA Journal, v51 n3, July 2010, p. 40.

Warning! Reading this book is certain to change the way you think about FM radio history, about Edwin Armstrong, and about Armstrong's best-known biographer, Lawrence Lessing.[1]

The research Frost did for his 2004 Ph.D. dissertation, "The Evolution of Frequency Modulation Radio, 1902–1940," puts him in good stead for writing this book. He seeks to reinterpret and sometimes rewrite the "iconic saga" of Armstrong now imbedded in many articles and books, including Lessing's. That story has Armstrong as a lone inventor, among the first to work with FM, whose patents of 1933 definitively and with received wisdom showed the world the several advantages of his frequency modulation system.

But we know that frequency modulation of telegraphy and telephony was patented in the United States in 1905 (patent number 785,803) by Cornelius Ehret. Never heard of Ehret? It's not surprising; Frost says that he "ranks among the most obscure inventors in the history of wireless" (p. 18). Valdemar Poulsen also used FM (reluctantly, Frost says), as shown in his 1902 Danish patent featuring an arc oscillator. In fact, Frost provides a table listing about two dozen FM patents granted before Armstrong's famous 1933 patents. Among them is a 1925 patent for an FM transmitter granted to Frank Conrad, a KDKA founder.

Any history of frequency modulation must make a stop, however briefly, at John R. Carson's classic paper, "Notes on the Theory of Modulation,"[2] in which Carson showed the unworkability of narrowband FM. (Carson coined the terms *amplitude modulation* and *frequency modulation* in this article). However, we sometimes see what we want or need to see, rather than what we are actually seeing. This was the case with many people who persistently misinterpreted Carson's article as concluding that *all* FM was unworkable. This misunderstanding supposedly delayed research in wide-band FM for years, but Frost states, ". . . blaming him for retarding the progress of frequency modulation generally is nonsensical." (p. 47).

All of this material builds up to the heart of Frost's book: his analysis of Armstrong's patents of 1933 and what he sees as Armstrong's understanding of them. The phrase "incremental technology" in Frost's subtitle and his interpretation of the word *serendipity* in the book provide good clues to understanding his contention that FM's history was not nearly as straightforward as currently thought, including Armstrong's part in it.

I do not wish to give away all of the surprises included in Frost's book. Suffice it to say that he provides plenty of background, research, thought and analysis for his ideas, and these ideas are sure to generate as many opinions as he has readers. But Frost's unique–I am tempted to write *groundbreaking*–book now becomes one whose ideas all future historians of FM must absorb.

It's a given that we sometimes invent things for the "wrong" reasons or we invent things whose properties and uses are not immediately apparent. To further whet the reader's appetite I conclude with a Frost quote that some Armstrong aficionados may find verging on the offensive. Frost states:

> *Before the issue date of the wideband FM patents, and for several years afterward, no one, not even Armstrong, completely understood the potential of what he had invented. He had based his patents on imaginative yet flawed theories of radio communication, and he therefore anticipated virtually none of the now well-known properties of modern FM radio, most notably its abilities to suppress static and reject interstation interference.* (p.77).

Now, doesn't that make you want to read the book?

1. Lessing, Lawrence, *Man of High Fidelity: Edwin Howard Armstrong, A Biography*, Philadelphia: Lippincott, 1956.

2. John Renshaw Carson, "Notes on the Theory of Modulation," *PIRE* 10 (Feb. 1922): 57-64.

Early History of Radio from Faraday to Marconi
G. R. M. Garratt and Susan Garratt, The Institution of Electrical
Engineers, 93 pages, 1994.
AWA Journal, v51 n1, January 2010, pp. 59-60.

At a quick glance this book might be taken for a brief summary of
known facts about half a dozen19th century radio people, but not so.
Garratt spent many years in researching and writing these pages. He's
a lucid writer who provides insights about his subjects that readers
may not find elsewhere. For example, see the exposition on page 12
of the thinking that Joseph Henry went through to arrive at his
conclusion that the discharge from a Leyden jar is oscillatory, i.e., it
repeatedly reverses direction.

The book includes material on Faraday, Maxwell, Hertz, Lodge and
Marconi. Ampère, David Hughes, Henry, Preece and Popov also
make brief appearances.

We learn that David E. Hughes–said to be the inventor of the
microphone–generated and detected electromagnetic waves years
before Hertz demonstrated them. There were just two problems– the
"amateur" Hughes did not entirely recognize his accomplishment,
and he was given poor advice from three "experts," William
Spottiswoode, Professor Huxley and Sir George Stokes.

In the Hertz chapter Garratt tries to show why Hertz vacillated so
long before making his decision whether to be an engineer or a
scientist, and why he initially declined to study electromagnetic
waves. Historical continuity is provided when Garratt describes
meeting Hertz's widow in 1938 and his daughter in 1958. Each
woman provided Garratt with original manuscripts of Hertz's
writings.

An entire book could be written about the personal and
professional relationships between the patient, "pure" scientist Oliver
Lodge and the single-minded entrepreneur Guglielmo Marconi. Until
then, the long Lodge chapter here provides some facts and opinions.
Lodge was vindicated in this story of competition when in 1909 his
tuning patent of 1897 was extended for seven years, thus
"embarrassing" the Marconi Company so much that an agreement

was entered into, with the Marconi Company paying the Lodge-Muirhead Syndicate 18,000 pounds (some writers state a different amount) for Lodge's tuning patent and other rights. The author states: "It would be difficult to conceive of a more complete vindication of Lodge's claims in this field."

Garratt somewhat reluctantly provides a four-page chapter on Aleksandr Popov, whom the Russians–or at least Russian spin artists–touted as the "true inventor of radio." But Garratt states that claims for Popov's priority "have a background of political propaganda, [and] have been disseminated so widely that it is necessary to take serious notice, even if only to demolish them." So much for Popov.

Garratt died before finishing his book; the last chapter, on Marconi, was completed by Garratt's daughter Susan. It contains details about Marconi's early radio demonstrations on Salisbury Plain and at Lavernock Point in England in 1896-1897, including the actual text of some messages that were transmitted. And there the book ends.

Early Radio Wave Detectors
Vivian J. Phillips, Peter Peregrinus Ltd, 223 pages, 1980.
AWA Journal, v51 n4, October 2010.

Most radio history buffs are familiar with the crystal/catwhisker detector popular during the 1920s. And some enthusiasts can recognize the cymoscope (pronounced *sigh-mo-scope*) used by Hertz to detect electromagnetic waves in the late 1880s. You can see a drawing of a cymoscope on page 49 in *S. Gernsback's Radio Encyclopedia,* and it's also part of the AWA logo, shown here. Between these two devices historically lie a myriad of instruments formerly used by science in radio wave detection.

Hertz learned much about electromagnetic waves by using the cymoscope, but its inefficiency coupled with the difficulty in tuning it drove the search for a more sensitive detector. Enter the coherer, the most well-known version of which consisted of a glass container holding metal filings in loose contact. When a radio wave from the receiver's antenna passed through the coherer it caused the filings to stick together, or cohere. This greatly lowered the resistance of the filings, allowing *one* telegraph signal to be received. Unfortunately the filings then remained stuck together and needed to be separated or decohered before they could process the next signal.

Oliver Lodge accomplished decohering by causing a small metal ball to tap lightly against the coherer's body, but Phillips shows there were other schemes to accomplish the same thing. These alternate methods resulted in detector instruments that, in most cases, would be unrecognizable as such to today's radio hobbyist.

Nor were coherers the only kind of pre-crystal detector in radio reception use. Electrolytic detectors depended on electrolytes, or conducting solutions, for their operation. In the chapter on these detectors, Phillips examines about a dozen instruments, including Reginald Fessenden's bubble electrolytic detector (British patent 4714 of 1907). He questions whether the device could actually work as Fessenden described, then quickly adds: "Still, bearing in mind Fessenden's massive reputation as an inventor and engineer, one must assume that the apparatus worked . . . "

In their turn, there follow chapters on magnetic detectors, thin-film and capillary detectors, and thermal detectors, each chapter containing photographs and line drawings, and careful explanations of how each device functioned (or was supposed to function). The advent of continuous wave broadcasting actually made the reception of Morse signals more difficult, for reasons explained in Chapter 8, "Tickers, tone-wheels and heterodynes."

The short chapter on miscellaneous detectors (those that didn't fit into any of the above categories) might be overlooked, save for the inclusion of "physiological detectors." These included the use of frog legs, then cat brains (dead cats at first, then cats "under the influence of an anaesthetic"). Here Phillips introduces the early 20th century work of A. F. Collins, who used human brains "within hours of the death of the physical body." The chapter includes a picture of Collins "listening to the 'cohesion of the human brain under the action of electric waves.'" The author cautions: "Readers of delicate sensibilities are advised to refrain from looking at Fig. 9.17," though by today's standards the photo may be considered viewable. One might idly wonder whether the brain of a longtime (though necessarily former) AWA member would produce cohesions markedly different from, say, the brain of an average Joe.

Finally, in the last chapter, we come to the familiar crystal/catwhisker combinations, Fleming's diode valve, and de Forest's audion, or triode. The book ends with a one-page table compiled in 1911 by S. M. Powell comparing efficiencies of various types of detectors.

Early Tube Development at GE
Henry Schroeder and William C. White, Tube Collectors
   Association, Inc., 38 pages, 2005.
AWA Journal, v47 n1, January 2006, p. 58.

We sometimes like to think that four-pin tubes from the 1920s just always were. Furthermore we like to think that they are exactly as they should be. Neither of these statements is true, of course, and Schroeder and White show why in these 38 pages written by two who were there at the time. A mini-history is given for each of the two-dozen or so tubes covered, and these often include information as to why the tube was made as it was.

The work of Irving Langmuir shines through this essay, and deservedly so. His series of inventions and insights ushered in the age of the thoriated-tungsten filament.

TCA (Tube Collector's Association) has added a few dozen tube photographs to this essay, plus an appendix on Albert Hull's AC-operated kenopliotron, which was in operation as early as 1923. This publication makes a nice companion to Robert Millard's *Collector's Vacuum Tube Handbook*.

Edison: Inventing the Century
Neil Baldwin, Hyperion, 531 pages, 1995.
Pittsburgh Oscillator, v10 n3, September 1995, p. 17. Reprinted in
OTB, v42 n3, August 2001, pp. 30-31.

This well-written and thoroughly researched biography contains little about Edison's radio activities, but radio enthusiasts will still find it interesting for its coverage of Edison's related work in DC generation and distribution, electric lighting, the phonograph, etc.

Having invented and promoted his system of direct current power distribution Edison was considerably opposed to the idea of alternating current distribution, particularly the Westinghouse AC system. Here Edison slipped into outright ghoulishness: in the courtyard of the West Orange laboratory, media-event public executions of animals by alternating current were conducted, escalating from dogs ("criticized because the weight of the animal killed was less than that of a man") to calves to horses. Edison's attorneys Eaton and Lewis suggested in all seriousness that recommended terms for this method (including "ampermort," "dynamort," and "electromort") were inaccurate in their derivation: "Electricide" seemed more fitting, expressing direct analogies in the English language with "homicide" and "suicide."

But perhaps the best choice of all was to invent a new verb, to "Westinghouse." "As Westinghouse's dynamo is going to be used for the purpose of executing criminals," Edison's cunning men reasoned, "why not give him the benefit of this fact in the minds of the public, and speak hereafter of a criminal as being 'westinghoused,' or (to use it as a noun) as having been condemned to the westinghouse, in the same way that Dr. Guillotine's name was forever immortalized in France?" (p. 202).

Edison may have had his blind spots; nevertheless, he was almost continually inventing, and patented over a thousand devices in his lifetime. Among these was his three-story cement house, which he designed and built in an attempt to provide cheap ($1,200) housing for the masses. A mold of 500 cast iron sections was bolted together on top of a concrete footing. Then, over a six-hour period, wet cement was poured into a funnel opening on the roof until the mold was filled. Four days later the mold was removed and the cement house required only windows and doors to be added. Ten such houses still stand today.

In his final years Edison's passion was the production of rubber from native American plants. He succeeded in producing 100 pounds of rubber from an acre of goldenrod.

It could easily be argued that Edison invented the first radio tube (U.S. patent 307,031 for an electrical indicator), but he found no practical use for the device in radio. And he very reluctantly permitted his sons to get into radio manufacturing, which they did by merging with Splitdorf Radio around 1929. Baldwin writes: "Despite the [Edison] brothers' alacrity, the new Edison Light-O-Matic radio in its big, ornately carved cabinet was born as a $1,000 dinosaur compared to the versatile 'little boxes' that were now invading American homes. The Edison radio was an eventual loser, surviving on the market for only eighteen months." (Pages 388-9; see also Alan Douglas's *Radio Manufacturers of the 1920's,* volume 3, pages 125-133).

Electronics in the West: The First Fifty Years
Jane Morgan, National Press Books, 194 pages, 1967.
AWA Journal, v49 n3, July 2008, p. 24.

Anyone who assumes uncritically that most U.S. electronics was discovered or invented in New York or Philadelphia can be enlightened by this book. It's an oldie all right, and in many ways technically naive (it was written for electronics neophytes), but this is compensated for by a rich array of biography. Here is the early development of electronics in the San Francisco Bay area, Palo Alto, Los Angeles, Berkeley, Stanford Research Institute, etc. In those places many people got their start or did major work. We have de Forest and Farnsworth, of course, and William Shockley (transistor), the Varian brothers (klystron), Ernest O. Lawrence (cyclotron), Bill Eitel and Jack McCullough (Eimac), and others.

Charles D. Herrold was broadcasting in San Jose in 1909. Peter Jensen (of Jensen speaker fame) and E. S. Pridham started the Magnavox Company in Oakland, California in 1917. Frederick Terman, the engineering textbook writer, taught at Stanford for more than 40 years. William R. Hewlett and David Packard supposedly flipped a coin to see whose name would appear first in their company. Jo Jennings realized that capacitors could carry higher voltages if they were in a vacuum. He invented the variable vacuum capacitor. Even Bing Crosby got into the act, buying 20 of the first Ampex tape recorders–Alexander **M. P**oniatoff produced **ex**cellent equipment.

Morgan's book was reviewed by Bruce Kelley in the March 1968 *Old Timer's Bulletin* when the book was available new for $4.95. Today it's available used for $9.95. Kelley advised his readers, "Yes, there were a couple mistakes but nothing to get excited about . . . ."

(See also Timothy Sturgeon, *How Silicon Valley Came To Be,* p. 82).

Empire of the Air: The Men Who Made Radio
Tom Lewis, HarperCollins, 421 pages, 1991.
Pittsburgh Oscillator, v6 n4, December 1991, p. 18.

We can't imagine anyone interested in radio history not being interested in this book.

Here is the non-technical history of radio from its beginnings through the 1950s, told largely through the concurrent biographies of three radio giants–Armstrong, de Forest and Sarnoff.

Armstrong and de Forest may have been mere rivals at the beginning of their relationship, but by the 1930s that rivalry had turned to hatred. It is extraordinary the means which each man took to have himself proclaimed the winner in the priority debate over who invented regeneration. The obstinate will to succeed which led the two men to make the great discoveries that they did was now used to pursue their legal battles with each other. Though legally de Forest was pronounced the winner by the U.S. Supreme Court on two different occasions, the scientific and engineering communities and many others continued to support Armstrong's claim.

In the late '40s and early '50s Armstrong was to relive this nightmare during his protracted legal battle with RCA and David Sarnoff when he sought redress for RCA's use of his wide-band FM inventions.

If there is a hero in this great bittersweet tale of radio invention and exploitation it must be Armstrong who, despite great public and professional adversity, continued throughout his adult life inventing and revolutionizing radio–regeneration in 1914, the superheterodyne in 1920, superregeneration in 1922, wide-band FM in the '30s, and FM radar and FM multiplexing, which lead directly to stereophonic transmission, among other things. Part of Armstrong's extraordinary technical success followed from his stubborn refusal to believe what "everyone" knew to be true, even when "everyone" apparently had mathematics on their side.

But Armstrong was a tragic hero, done in by his pride, which was manifested by his seeming inability to compromise when he thought he had been wronged. Several times in his life he refused to compromise when by doing so he could have walked away a successful and wealthy man. But it was not in his nature to do so, and in the middle of his court battle with RCA he committed suicide in 1954.

When Sarnoff, then president of RCA, heard the news of Armstrong's death, he said, "I did not kill Armstrong." The statement is so obviously true that one immediately wonders why Sarnoff felt impelled to make it.

De Forest's reaction to Armstrong's death was somewhat different. In a letter to a friend he wrote:

> *His death was indeed most lamentable. I have always given him full credit for his introduction of FM, but have always taken the keenest delight in having beaten him so thoroughly on the feed-back question . . . . Well, after all, Armstrong has gone and I am alive, well and happy, and hope to live for many years more. What a contrast!*

Do not dismiss the "Sources and Notes" section of this book as important only to librarians and researchers. There is relevant information here that does not appear in the main text, including the bizarre theory of Senator Joseph McCarthy that Armstrong had been murdered when he refused to divulge information about his FM radar research to the Communists.

This popular study–well written, extensively researched, and highly readable–will change the way you think about its three protagonists and radio history in general.

The book has been made into a documentary film of the same title by the author and Ken Burns (of *The Civil War* fame).

Encyclopedia of Radio
Christopher Sterling, ed., Fitzroy Dearborn, 1,650 pages, 2003.
OTB, v45 n2, May 2004, pp. 25-26.

This beautiful set is enough to get this old librarian's heart racing. Think of it–1,650 pages of radio stuff! Reason demands I start by reading the editor's introduction, but I don't. Instead I flip open volume two at random and start turning pages. I notice first the two-column-per-page layout (typical for multi-volume encyclopedias), pleasing typeface and wide margins, and paper of obviously high quality. I skip through a few articles and finally settle on "Italy," the first paragraph of which tells me much I didn't know:

> *Italy, the country where radio was invented, is home to one of the most advanced and diversified of the world's radio systems. The distinguishing feature of Italian radio is its 3,000 different stations, a figure that places Italy second only to the United States in the total number of signals available. With an average of one radio set for each of its 60 million people, Italy has one of the world's highest levels of radio penetration. Generating more than $400 million in annual revenues, Italy's commercial radio sector is the seventh largest in the world.*

General encyclopedias by definition attempt to cover all knowledge; specialized encyclopedias attempt to cover all knowledge on a particular subject. Of course, neither can. Humanity has been producing information for millennia and, recently, at such a rate that the best any encyclopedia can now do amounts to an organized summary or outline. For example, the present work is international in scope, and this presents a considerable challenge to the editors. Most of the 20 most populous nations receive comprehensive over-views, with the exceptions of Ethiopia, Egypt and Nigeria. (Surprisingly, Iran information is indexed under "Persia"). The editor explains:

Our approach is international in scope, though with a strong emphasis on the United States and a secondary emphasis on key English-speaking nations including Britain and Canada (volumes such as these could be assembled for many other countries). Other regional entries provide at least brief comparative comment on most other systems of and approaches to radio broadcasting.

Note carefully the title: *Encyclopedia of Radio*, not *Radios*. So hardware enthusiasts may be disappointed. There are no articles about particular radios (though there is an article on receivers) and no pictures of radios. In fact, the only photos are black-and-white portraits accompanying biographical articles. There are about 200 such articles and many of our heroes are included–Armstrong, de Forest, Fessenden, Marconi, etc. A few of our non-heroes are also here–Axis Sally, John R. Brinkley (the "goat gland" doctor), Tokyo Rose, etc. Each biographical article is followed up with a summary which includes important dates and achievements, awards, and publications. Other prominent entry types are radio programs and personalities, organizations, and station histories. An extensive 44-page analytical index closes the work.

Even though it is published at the beginning of the age of electronic information, this old-fashioned paper-and-ink encyclopedia will remain a landmark in radio literature for years. Printed at substantial expense and aimed at the library market, it's way too expensive for the average radio hobbyist, though I see a copy of the set listed for $149.00 at:

www.strandbooks.com

("eight miles of books," many at half price or less). Even at that price, I suspect it won't be there long.

Ernst Fredrik Werner Alexanderson: One Individual's Life and Contributions to Electrical Science in the First Half of the 20th Century
Bengt V. Nilsson, available from *Antique Radio Classified,* 71
  pages, 2006.
AWA Journal, v48 n1, January 2007, p. 17.

This book is not to be taken as a full-fledged biography of Alexanderson. For that, see the Brittain book reviewed above (page 2). Instead, Nilsson's book presents a series of essays or vignettes, each dealing with some aspect of his life and career at General Electric and RCA. The essays cover only the highlights of Alexanderson's interests, including the high-frequency alternator that started his career, his relations with Steinmetz, Langmuir and many other engineers, and his work with facsimile, monochrome and color television, thyratrons and DC power transmission, etc.

Alexanderson also found time to act as a sort of unofficial emissary for Sweden, his country of birth, by entertaining and consulting with many Swedish engineers who visited the United States and General Electric.

The book includes a list of Alexanderson's patents, and reproduces a flyer printed by the GE publicity department showing that Alexanderson had produced "an invention every seven weeks" up to that point. His patent count, though certainly impressive, was eclipsed by that of his colleague John H. Hammond, Jr., by George Westinghouse earlier, and most certainly by that master patenting machine Thomas Edison.

About a hundred photos, drawings and charts enhance the text.

Experimental Researches in Electricity
Michael Faraday, 1839-1855, reprinted by Green Lion Press in
  three volumes, 2000.
OTB, v43 n4, November 2002, p. 45.

  Modern facsimile reprints of out-of-print science classics are
always welcome. Though the findings reported in these volumes
have long since been absorbed into the literature of science and
included in countless textbooks, historians of science in particular
will be happy with this opportunity to delve into Faraday's writings
just as he put them down in mid nineteenth century.

*Faraday's Experimental Researches in Electricity: Guide to a First
Reading*
Michael Faraday and Howard Fisher, Green Lion Press, 617 pages,
  2001.
OTB, v43 n4, November 2002, p. 45.

If 1,500 pages in three volumes of original Faraday is a little more
than you might need, Howard Fisher presents a workable
compromise with this volume. Fisher has studied Faraday "in both
laboratory and library" for over 25 years. Here he selects for reprint
about a third of the items from the work above. About a fourth of this
volume consists of Fisher's own insightful and thoroughly-informed
commentary on Faraday's text.

Fabrication d'une Lampe Triode
*http://blog.makezine.com/archive/2008/01/make_your_own_*
  vaccum_tube.html (Note: vaccum, *not* vacuum).
AWA Journal, v49 n2, April 2008, p. 34.

Lately I've had my consciousness raised yet again. I try not to say, "I built that radio [or stereo, hi-fi, etc.] myself." Now I try to say, "I assembled that radio from a kit," for in truth I have never built a radio by determining the parts, designing the layout, etching a circuit board, etc. All my electronic life I've also steadfastly avoided bending, drilling or punching a chassis. Just too much work for me.

Now, along comes this guy (in France?) who takes the whole thing one step further–he builds his own radio tubes. This can only be described as *Do It Yourself* writ large.

The seventeen-minute movie is spell-binding. To show how easy it is to make a tube–or perhaps just to show *off*–the builder takes out his pocket knife, opens from it a baby pair of scissors and uses them to cut the plate of the tube from a thin sheet of unidentified metal.

Then the fun begins. To the endless repetition of the song "The Man I Love," the tube-maker takes us through each step, including glass cutting, shaping and molding, tack welding, drawing a vacuum, etc. It all appears to be done in someone's basement workshop, or in a small metal- and glass-working facility.

The final result is a working triode, completely hand crafted, now operating in a one-tube radio. All fascinating stuff. Get yourself to an internet computer with a fast connection and I can pretty much guarantee you'll be entertained and impressed.

Fireside Politics: Radio and Political Culture in the United States, 1920-1940
Douglas B. Craig, Johns Hopkins University Press, 362 pages, 2000.
OTB, v42 n2, May 2001, pp. 62-63.

This is no hobby book. It is an academic study, serious and unexciting in style but so richly-detailed and analytical as to set a standard. The author, a senior lecturer in history at the Australian National University, knows his subject well; he previously wrote *After Wilson: The Struggle for the Democratic Party, 1920-1934.*

The book's broad subject is the influence of radio on politics–and politics on radio–during the inter-war years. The thoroughness of the study is evident in a glance at its subheadings: *advertising, formation of networks, government regulation, the Federal Radio Commission, the Communications Act of 1934, the FCC, campaigns and political programming, audience measurement, voter behavior,* etc. A study of radio, citizenship and good taste forms a long coda to the work.

Strangely enough, television might get in the way of the full appreciation of this book. Young readers may wonder what all the fuss was about, not realizing that radio was the only game in town during the '20s and '30s. With no competition from television, radio audiences were huge. In the twenty-year period covered by the book the percentage of American homes equipped with radio went from essentially zero to 80%. By contrast, during the same period homes equipped with telephones remained steady at about 40%.

Everyone listened to radio. I don't mean they had the radio 'on'; I mean they listened to it. Hence, radio could actually affect, say, the way people voted or what they bought–or thought. One of the social phenomena evident throughout the book is that of people struggling to understand, manipulate or exploit the new medium of radio, not always with the expected or hoped-for results. For example, Craig takes four pages to describe the 1928 Hoover-Smith national contest– "the first true radio campaign"–then concludes with, "Amidst the general celebration of the newfound power of radio, nobody seemed to notice that the losing party had spent more on radio than the winner."

I'm glad this book is hardbound, for it will be used for many years in libraries and reference collections. Typeface, layout and physical organization are all in keeping with the scholarly nature of the book. (Kudos to the savvy designer who put a stylized radio dial at the head of each chapter). Other visual features include charts, maps, and reproductions of contemporary cartoons. Readers who saw the movie *A Face in the Crowd* can see it succinctly reprised in the cartoon on page 169.

Like any good scholar, Craig not only tells us what happened and why, but also where he got his information. So the book includes a twenty-page bibliography listing, among many other sources, about 50 archive collections. These, together with 43 pages of notes and a detailed index make *Fireside Politics* a virtual *vade mecum.*

From Immigrant to Inventor
Michael Pupin, Scribner's sons, 396 pages, 1923.
OTB, v42 n4, November 2001, pp. 29-30.

Biography merely permits us to snoop into other people's lives; autobiography invites us to. Michael Pupin's autobiography is no exception, but modern readers are advised to consider that Pupin was born to Serbian peasants in the middle of the 19th century in Idvor, a village that was part of Austria-Hungary and so small it "cannot be found on any map." Pupin spent the first 14 years of his life there, and the rest of his life was thoroughly informed by his childhood experiences.

Thus Pupin's book is not what we would today call a "quick read." It is slow and ponderous, written with a 19th century outlook. It is the kind of book one might set aside to read after one retires. Pupin often employs broad metaphors presented in complicated sentences and paragraphs so long they won't fit on a single page. Nevertheless, his autobiography earned a Pulitzer Prize in 1924.

Don't look here for details of Pupin's inventions or research. Social and philosophical commentary makes up a good portion of the text. Here, for example, is Pupin on racism: "Racial antipathy is one of the saddest of psychic derangements; and, although it is a repulsive product of modern nationalism, the world does less than nothing to get rid of its insidious poisons. European civilization is being destroyed by it." Shall we add to his list of talents the ability to predict?

Pupin continually harks back to his days as a young herdsman in Idvor, charged with the safety of cattle during the night when they were vulnerable to theft by neighboring villagers. He and his fellow herdsmen communicated secretly by thrusting a knife into the ground and tapping on the knife handle. Pupin quickly learned that dry, hard soil transmitted messages much farther than wet or loose soil.

He claimed later in life to have used this knowledge when in 1899 he (and/or others) invented the loading coil that permitted long-distance wired telephony. His patent came none too soon, as George Campbell and Edwin Colpitts of American Bell were working in the same area. Telephone lines that incorporated the invention were said to have been "pupinized" and were used worldwide.

It is possible to carry allegiance to a metaphor too far, however, and thereby draw the wrong conclusions. Here is Pupin writing in 1923:

> *Every now and then we are told that wireless signals might be sent some day to the planet Mars. The judgment of a former herdsman of Idvor considers these suggestions unscientific for the simple reason that we cannot get a ground on the planet Mars and, therefore, cannot take it into close partnership with our Hertzian oscillators. Without that partnership there is no prospect of covering great distances.*

At the age of 15 Pupin traveled alone to America, arriving in New York City with five cents in his pocket and a great need to convince immigration authorities that he should not be deported. He did that by telling them of his admiration for Abraham Lincoln and Benjamin Franklin. He got a job as a mule driver, and immediately set about learning English. Later he taught himself French and German. (A recent PBS documentary on the life of another Serbian, Nikola Tesla, has the young Tesla arriving in New York City with *four* cents in his pocket. There are other parallels between the early careers of Pupin and Tesla, but that is another story).

Throughout his career Pupin sought an answer to the question *What is light?* When his professors at Columbia College (now Columbia University) in New York City could not provide an adequate answer, he traveled to Cambridge University, meaning to study with his idol, James Clerk Maxwell, only to find upon his arrival that Maxwell had died four years previously. Still seeking answers to his question he next traveled to the University of Berlin. There his Ph.D. advisor was the great physicist Hermann von Helmholtz. Pupin also idolized Helmholtz, and refers admiringly to him throughout his book by his German title, "Excellenz von Helmholtz."

Pupin received 34 patents for his work in the practical aspects of telecommunications. (Six of them were issued jointly to Pupin and his pupil, the soon-to-be successful–though not famous–Edwin H. Armstrong). Nevertheless his book can be seen as an extended plea for the virtues of pure research over applied research. He wrote, "The worship of the eternal truth and the burning desire to seek an ever-broadening revelation of it constitute the mental attitude which I call 'idealism in science'."

Great Feuds in Technology: Ten of the Liveliest Disputes Ever
Hal Hellman, John Wiley & Sons, Inc., 248 pages, 2004.
AWA Journal, v47 n1, January 2006, p. 58.

I found this book captivating, and I assume that anyone with a lively interest in the lively history of technology would also. In ten concise chapters Hellman dispassionately lays out the case made for each player in priority disputes, patent disputes, and just plain old personality conflicts. Readers can root for their favorite opponent (or victim) in essays on the Luddites *v.* the Industrial Revolution, Davy *v.* Stephenson (miner's safety lamp), Henry Ford *v.* George Selden and the Association of Licensed Automobile Manufacturers (ALAM), the Wright brothers *v.* Curtiss *et alia*, Rickover *v.* Zumwalt "and Just about Everyone Else." There are also chapters on major disgruntlements in the human genome and biotech worlds.

But most AWAers will want to read the chapter on Morse *v.* Charles Jackson and Joseph Henry in the fight over the invention of the electric telegraph. Morse may have won this battle, but it was only at the cost of a years-long struggle which generated bitterness, particularly between Morse and Henry.

There is an essay on Edison *v.* Westinghouse, or the "AC/DC War." Many animals lost their lives as a result of this war when two of Edison's associates–Harold P. Brown and Arthur E. Kennelly (of Kennelly-Heaviside layer fame)–demonstrated the dangerous properties of AC by staging public executions. The victims of deliberate electrocution escalated from dogs and cats to horses and, ultimately, to human beings (See Mark Essig, *Edison and the Electric Chair*, 2003). Edison, however, decidedly did *not* win the AC/DC War, though Hellman states: "Direct current is in use today for equipment that may have been installed as late as the 1920s and continues to operate."

Lastly, as a prime example of winning the battle but losing the war, there are 20 pages on the battle between Sarnoff and Farnsworth, whom Hellman calls, "The Fathers of Television." Was Farnsworth robbed by RCA in this great priority fight? You be the judge. (Many argue for John L. Baird as television's inventor, but he gets short shrift in this essay).

Hellman is also the author of *Great Feuds in Medicine*.

Great Physicists: The Life and Times of Leading Physicists From Galileo to Hawking
Will H. Cropper, Oxford University Press, 500 pages, 2001.
OTB, v44 n1, February 2003, p. 35.

Readers of *The OTB* will probably be most interested in the Faraday and Maxwell biographies here, but the book as a whole is rewarding reading. For each of 30 great physicists Cropper attempts to summarize the most important achievements. He also attempts to present each biographee as a human being. This results in the inclusion of many unprofound though interesting facts, made all the more so when contrasted against the weightiness of the book's main subject.

So in the chapter on Hawking, for example, we find out that during WW II bombing "German *Luftwaffe* agreed to spare Oxford and Cambridge if the Royal Air Force would do the same for Heidelberg and Göttingen." And in the essay on Fermi we learn that Italian pencils have no erasers. What a testament to self-confidence!

Includes Bohr, Boltzmann, Carnot, Chandrasekhar, Clausius, Curie, de Broglie, Dirac, Einstein, Faraday, Fermi, Feynman, Galilei, Gell-Mann, Gibbs, Hawking, Heisenberg, Helmholtz, Hubble, Joule, Maxwell, Mayer, Meitner, Nernst, Newton, Pauli, Planck, Rutherford, Schrödinger and Thomson.

Heinrich Hertz: A Short Life
Charles Susskind, San Francisco Press, Inc., 194 pages, 1995.
AWA Journal, v47 n2, April 2006, p. 24.

It's too bad that this is a short life of Hertz, for reading it immediately makes one want more. But then Hertz's life was short. He died of septicemia in 1894 at age 36.

Most of us know Hertz merely as the guy who verified the existence of electromagnetic waves. This book goes way beyond that to show us Hertz the man. We know much about him through his own book and magazine publications, but also through his diaries and correspondence, much of which has been preserved. Hertz's daughter Mathilda, who died in England in 1975, was, nevertheless, very influential in the publication of this book.

How did Hertz, the young German scientist-to-be, prepare himself (unknowingly) to make his great 1887-88 discovery? The youthful Hertz studied physics, advanced mathematics, oceanography, elasticity, French, English–and Arabic! Later his interests became much more specific: "hydrodynamic instability, observations of electromagnetic induction, diffusion of light in electrolytes, stability in thin liquid films, surface properties of bubbles and of falling raindrops, optical interference phenomena." Remember studying those subjects in high school?

It is quite fair to say Hertz was a genius. Helmholtz wanted Hertz to investigate electromagnetism as a subject for his dissertation, but Hertz demurred, thinking the work would be too extensive and difficult. Instead he chose a topic he thought he could finish in a year. He finished the work in less than three months. Only years later did Hertz complete the experimental and theoretical work that conclusively verified Maxwell's theory of electromagnetic wave propagation through space, simultaneously disproving Weber's theory of "action at a distance" in the ether.

Being a scientist rather than an inventor, Hertz let others exploit and profit from his 1887 rediscovery[1] of the photoelectric effect. Similarly, Hertz did nothing to develop the applications of electromagnetic waves, though as Susskind insists, Hertz certainly realized that they had great potential. As a scientist, however, Hertz did publish careful accounts of his investigation of electromagnetic waves, and his work was noted with keen interest by a young inventor in Italy named Marconi.

1. See G. W. A. Dummer, *Electronic Inventions and Discoveries,* 4th edition, page 82.

Hi Hi: A Collection of Ham Radio Cartoons
Dick Sylvan, Sylvan Design Associates, Ltd., 111 pages, 2005.
AWA Journal, v47 n3, July 2006, p. 41. Reprinted in *Pittsburgh Oscillator*, v21 n3, September 2006, p. 16.

The book's title derives from amateur radio jargon used to indicate laughter. Try it yourself: four dots (or dits), a pause, then two dots. Repeat this sequence until you start laughing. (See *http://www.ac6v. com/jargon.htm* for more ham jargon, abbreviations and terminology).

Sylvan's six decades of ham experience (that's a whole lot of dits and dahs) inform each of these cartoons, many originally produced for *K9YA Telegraph*, an international amateur radio e-Zine. About 100 highly-detailed black-and-white cartoons running the gamut from ham lingo educational to plain cornball silly. For the ham in all of us.

Historical Dictionary of American Radio
Donald G. Godfrey and Frederic A. Leigh, Greenwood Press, 520
  pages, 1998.
Pittsburgh Oscillator, v14 n2, June 1999, p. 6. Reprinted in *OTB*,
v44 n1, February 2003, pp. 35-36.

The word "dictionary" in the title is somewhat of a misnomer.
Although entries in the book are alphabetically arranged and many
radio terms are defined, the book is, in fact, a single-volume
encyclopedia of American radio. Entries range from a few lines to a
few pages in length. Though the period covered extends from radio's
beginnings, many of the articles describe current radio features and
practices.

Hardware enthusiasts will be disappointed. The book emphasizes
the programming side of radio, and there are no illustrations,
graphics or charts. Nevertheless, nothing has appeared yet that comes
even close to what this reference volume does for American radio
and its history. Sidney Gernsback's well-known 1927 *Radio
Encyclopedia* (which does emphasize hardware) makes a nice
historical companion.

Interspersed with concise definitions of radio terms (e.g., hot clock,
narrowcasting, payola-plugola, diary, duopoly, topless radio, zoo
format, etc.) are more extensive articles on general subjects and
historical processes in radio, such as radio legislation, racial issues,
propaganda, the fairness doctrine, public radio, the press-radio war,
etc.

Biographies appear for such diverse on-air personalities as
Wolfman Jack, FDR, Don Imus, and the Lone Ranger. Behind-the-
scenes people (Frank Conrad, Mahlon Loomis, Marshall McLuhan,
Marconi, Sarnoff, Owen Young and many others) are also present.
Roy Rogers gets a one page entry; Gene Autry gets none (though
information on him can be found by using the index). Radar gets no
article in the book, no entry in the index, and is not mentioned in the
article on radio and WW II.

Company and organization histories are included–the BBC, CBS,
FCC, NPR, Pacifica Foundation, Press-Radio Bureau, RKO, United

Fruit Company, etc.

The book includes an extensive 27-page bibliography of sources. Many of the articles have bibliographies attached, and some of those include World Wide Web addresses. This feature will be seen more and more in reference and other books, but readers are cautioned about the often-short-lived nature of Web addresses and information. A handy but limited eight-page radio chronology (1837-1997) is included in the book's front matter.

About 100 authors contributed entries. Surprisingly, Alan Douglas is not among them although his works are referred to throughout the book. Contributors worked independently, and this sometimes results in repetition of information, but the editors have done a good job of controlling the pace, tone and style of the book as a whole. It can be enjoyably read through from cover to cover.

Historical Dictionary of American Radio is enthusiastically recommended. It will see years of use as a starting point for researchers, librarians, radio historians and the general public. We hope someone is now working on the *Historical Dictionary of American Television*, and that it will be of equal quality.

Historical Sketch of Henry's Contribution to the Electro-Magnetic Telegraph
William B. Taylor, USGPO, 99 pages, 1879.
Pittsburgh Oscillator, v15 n2, June 2000, pp. 5-6. Reprinted in *OTB*, v43 n1, February 2002, pp. 64-5.

When did Samuel F. B. Morse invent the electric telegraph, and who invented it before him? One can learn a lot trying to answer this cock-eyed question. Anyone who seriously tries to do so will sooner or later arrive at Taylor's fascinating publication with the unwieldy title of *An Historical Sketch of Henry's Contribution to the Electro-Magnetic Telegraph: with an account of the origin and development of Prof Morse's invention.*

In 100 detailed pages Taylor recounts the history of the telegraph, beginning by quoting Robert Sabine: "The electric telegraph had properly speaking, *no inventor*. It grew up little by little, each inventor adding his little to advance it toward perfection."[1]

For the telegraph to be invented, it had first to be imagined. Here Taylor quotes a charming 17th century dreamer:

> *Among the numerous flights of imagination by which genius has frequently anticipated the achievements of her more deliberate and cautious sister–earth-walking reason, none is perhaps more striking than the romantic conception by Famianus Strada, of Rome, in the early part of the seventeenth century, of an intercourse maintained between separated friends by means of two sympathetic magnetic compasses, whereby the indications on the dial given by one, were instantly made visible to the other.*[2]

Taylor goes on to describe the many efforts that preceded Morse's, Wheatstone's and Cooke's commercially successful telegraphs. He begins with Lesage's telegraph of 1774:

> *The first electric telegraph of which there is record, is that established at Geneva by Georges-Louis Lesage. The line*

consisted of 24 insulated wires for the alphabet, each terminating in a pith-ball electroscope duly lettered, for indicating by its excitement the succession forming the words and sentences given by the operator, who employed at the transmitting station a manual conductor from an electrical machine.

Taylor finally gets around to describing Morse's efforts of half a century later, but first goes into detail about the telegraphs of M. Lambent (1787), M. Reiser (1794), Tiberius Cavallo (1795), D. F. Salva (1798), Dr. Samuel Thomas von Soemmering (1808), Dr. John R. Coxe (1816), Francis Ronalds (1816), Andre Ampere (1820), Paul Schilling (1823), Peter Barlow (1824), Harrison Gray Dyar of Long Island, New York (1828), Victor Triboallet de Saint Amand (1828), Gustav Fechner (1829), Dr. William Ritchie (1830), Carl Gauss and Wilhelm Weber (1833), and Professor C. A. Steinheil (1836). Only then are Morse's later "relay" and "receiving" circuits covered. Along the way Taylor credits Galvani, Volta, Oersted, Sturgeon and, most of all, Joseph Henry for their contributions to the science of telegraphy.

In addition to providing lots of detail and insight into the background of telegraphy–including the famous "disagreement" between Morse and Henry–the booklet unintentionally gives a thorough lesson in the potential pitfalls that await the naive researcher who sets out to determine a "first" in the history of invention.

1. *The Electric Telegraph*. Robert Sabine. London, 1867, part I, chap. iv, sect. 39, p. 40.
2. *Prolusiones Academicae*. F. Strada, quarto, Rome, 1617, lib. ii, prolusio 6.

History & Evolution of the Microphone
Bob Paquette, privately published, 842 pages, date?
OTB, v44 n4, November 2003, p. 32.

Weighing in at six pounds, this gargantuan book has many of the earmarks of a self-published labor of love: each copy is hand numbered, the pages are oversize and there are many of them, there is uneven print quality ranging from very poor to very good, much of the text is reproduced directly from other sources, and the book was assembled over a period of very many years. Anyone with an interest in–or obsession with–microphones will want to spend time with this book.

The majority of the book consists of reproductions (some of which are unattributed or incomplete) of such diverse document types as bulletins, catalogs, internal memos, manufacturer codes and specs, instruction and technical data sheets, drawings, newspaper and periodical articles, book chapters, advertisements, monochrome and color photos (some of these are downright beautiful), patent specifications, museum and corporate publications, government documents, proceedings, personal correspondence, etc.

The wide diversity of sources here makes the book cry out for a comprehensive index including, at minimum, all makes and models and all personal names. Had that been included, the book could have become a standard in the field; as it stands the huge amount of information here will nevertheless attract researchers who will use it as a source book or starting place for information not easily found elsewhere. For the microphone, there is nothing else like this book available.

History of Telegraphy
Ken Beauchamp, Institution of Electrical Engineers, 413 pages, 2001.
OTB, v43 n1, February 2002, pp. 63-64.

It is sad that Beauchamp (1923-1999) did not live to see his beautiful book published. But he did complete the text and choose the illustrations for this detailed history of the most useful device invented in early information technology. Beauchamp held a Ph.D. in signal processing and retired as the Director of Computing at the University of Lancaster (UK) in 1985.

To establish a baseline Beauchamp begins with a short chapter on telegraphy's visual predecessor, semaphores. This in itself is a fascinating subject. Russian Tsar Nicholas I established a system of semaphores stretching from St. Petersburg to Warsaw. It was staffed by 1,320 personnel and its 220 stations were strung across 600 miles. Interestingly, both semaphore systems and wired telegraphy, including submarine cable, were used in the Crimean War.

The remainder of the book is divided into two equal parts–terrestrial (wired) and aerial (wireless) telegraphy. Beauchamp describes some early telegraphy schemes that never caught on. One was Samuel Sommerring's military electrochemical telegraph. The receiver consisted of a cable terminating in 35 gold points submerged in a tank of water. Each point represented a number or letter. The device was read when electrolysis produced a stream of bubbles above the appropriate point.

One of the "subplots" of wired telegraphy is submarine cable. Some interesting wrinkles of submarine telegraphy revealed by the book are:

• The corrosive nature of sea water required the use of a particular type of insulation for submarine cables. This turned out to be gutta-percha, the same stuff that was packed into your tooth the last time you had a root canal procedure. The substance comes from the percha trees of southeast Asia. The use of gutta-percha by the submarine cable industry resulted in the deforestation of large areas of southern Malaya and Singapore in the mid 1800s.

75

• Because of the electrical characteristics of a long submarine cable (it can be seen as a large capacitor), a dot and a dash used in "cable code" are of the same time duration, the distinction between dots and dashes being made instead by a change of signal polarity.

• In time of war, cable ships repaired submarine cables deliberately damaged by the enemy, or better yet, spliced enemy cables into the cables of their own nation, thus neatly turning them from enemy cables into friendly cables.

• The 6,600 mile long cable between Australia and Canada was completed in 1902, years before electronic amplification was available. Fears that the cable was too long for practical use–that "No signals would come out at the other end!"–proved to be unwarranted.

Of equal interest are these wrinkles from the world of wireless telegraphy:

• Installation of telegraphy in aircraft during World War I increased the target accuracy of artillery sixfold.

• Marconi proposed a long-wave wireless system, the World-Wide Imperial Wireless Scheme, to link the entire British Empire in the early 1920s. Marconi's employee Charles S. Franklin was experimenting at the time with short waves. In the nick of time, Marconi changed his proposal to incorporate short waves and the resulting system was highly successful.

Evolution has operated without purpose for millennia to produce the living world we know today. Has telegraphy and 19th century technology in a similar manner produced the information technology world we know today? Beauchamp ends his book thus:

> The remarkable developments in 'communication at a distance,' commencing with the invention of the telegraph in the eighteenth century, can clearly be recognized as leading directly to the present equally remarkable growth in digital global communications–the wheel of technology has indeed turned full circle!

History of Television, 1942-2000
Albert Abramson, McFarland & Company, 309 pages, 2003.
OTB, v45 n2, May 2004, p. 25.

Purists will complain that Abramson's volume is misnamed, that it should not be called a history because it "merely" tells what happened, not why. But no matter. What the book may lack in interpretation is more than compensated for by the profuse enumeration of television facts. There are many thousands of these, all carefully documented. The footnotes alone number over 1,600, and many of them include additional facts not shown in the main text.

The extensive coverage quickly reminds us that those who think of television as only the box in the living room are severely limited in their perceptions. Abramson covers every major television equipment and technology development in the past half century, including a great deal on electronic moving picture recording and playback.

In the chapter "Death of RCA, or the G.E. Massacre," Abramson flatly describes GE's dismantling of RCA's radio research center in the late 1980s as "a crime." (Lots of interpretation in *that* statement). For a kinder, gentler view of this event see the Magoun book reviewed on page 34.

Abramson's earlier volume covers the years 1880 to 1941. Together these two books comprise a cornucopia of television facts. As such, the set is likely to be cataloged into reference collections in libraries. Librarians and researchers may be disappointed that the index to the book was not more robust. With literally dozens of facts listed on each of the 260-some main pages of text, a more detailed index was in order for this book. For example, some people important enough to have their photographs in the book are nevertheless not listed in the index.

This two-volume history of television represents an enormous amount of work accomplished essentially by one person. Compare with the *Encyclopedia of Radio* above, which was accomplished only through the labors of some 240 contributors.

Abramson's *History of Television, 1880-1941* is reviewed on page

37 of the November 1988 *OTB*. For Richard Brewster's interview of Abramson see pages 38-39 of the February 1997 *OTB*.

History of Wireless
Tapan K. Sarkar and others, Wiley-Interscience, 655 pages, 2006.
AWA Review, v50 n1, January 2009, p. 36.

Do not confuse this book with the Brodsky book of the same title reviewed on page 80. Brodsky's book is a popular review of the subject's technology; Sarkar presents a scientific treatment, written by many authors. It is not aimed at the radio hobbyist.

Following an introduction Sarkar and others provide a 100+ page chronology of wireless (1800 to 1990) which will delight any radio history fact collector. The caveat is that the chronology is essentially a compilation of unverified facts from about 100 secondary sources. Hence the incorrect statement that "[Édouard] Branly received the Nobel Prize in Physics in 1921"[1] is carried over into the chronology. (I can go on about this matter, for by an ironic twist of poetic injustice, I too repeated this very same error in one of my books.) In his extensive preface Sarkar quotes Pope Leo XIII (1810-1903): "The first law of history is to dread uttering a falsehood . . . ." Wise words indeed. Remember them the next time Wiley-Interscience vets your manuscript endlessly.

Entire in-depth chapters are then provided on Heaviside, Fleming, Tesla, Maxwell's equations, wireless before Marconi, etc. What is not expected are the thorough chapters covering waveguides, phased array antennas, J. C. Bose's work, wireless telegraphy in South Africa, antenna development in Japan, and the development of Soviet quasioptics. Quasioptics, for those of us who may not know, deals with radio at near-millimeter and sub-millimeter wavelengths. Want to know more? This chapter references 251 related publications!

Most chapters in the book contain diagrams, drawings and historical photographs and all end with detailed bibliographies, so each chapter can be used as an introduction to its subject or as a guide to further research. A comprehensive index to the entire collection of essays closes the book.

1. See *http://nobelprizes.com*

History of Wireless: How Creative Minds Produced Technology for the Masses
Ira Brodsky, Telescope Books, 250 pages, 2008.
AWA Journal, v49 n4 , October 2008, p. 40.

This is one of those books that I didn't want to like because the front cover pictures an unidentified, sleek, ultra-modern, hand-held wireless device (a cell phone?) and I gave up my cell phone a year ago because of disuse. But of course *The History of Wireless* isn't about just modern-day personal wireless. The book attempts to tell the long story of how civilization arrived at that sleek device.

And what a story Brodsky tells us. In his desire to be comprehensive, he begins by mentioning Thales of Miletus who, in sixth century BC, was unknowingly involved in the history of wireless when he described static electricity. (In fact, finding a starting point for the history of wireless is not a clear-cut task). Brodsky then settles in at the Luigi Galvani-Alessandro Volta controversy over animal electricity in 18th century Italy. Volta appears to have won the dispute, inventing the battery in the process.

There follow chapters of the author's interpretations of by now well-known discoveries, inventions and personalities, taking us up to the contemporary scene. "Wave Makers" explores the differences in approach used by Oliver Lodge and Heinrich Hertz in their investigations of the ether and electromagnetic waves. "Morse and Bell" provides a "grand introduction to the trials and tribulations of inventors". "The Father of Broadcasting"–no, it's not Conrad, Marconi, Fessenden or de Forest–is a 19-page biography of the Russian immigrant David Sarnoff. "Going Mobile" provides a mini-history of Paul Galvin's successful automobile after-market company, a.k.a. Motorola.

Brodsky worked in "high-tech industry" for 30 years, and it is for that time period that his history shines. His descriptions of the details of current cell phone and digital wireless are fact-packed, introducing many new, sometimes complicated, techniques. For example, we techies all learned at our mothers' knees that multi-path propagation causes fading and, in television, ghosts, and is to be avoided at all costs. But MIMO (multiple input/multiple output) wireless systems don't just tolerate multi-path propagation, they require it and in so doing greatly increase the system's carrying capacity.

New technologies sometime mandate new vocabularies. We who pretty much got by with AM/FM, AC/DC and RF/IF/AF tend to slow down when we encounter CEPT, TDMA, CDMA, D-AMPS, GSM, CTIA, etc. Or how about this sentence: 'Consequently, Airgo Networks created its "True MIMO" technology around MIMO-OFDM'? These are all from Brodsky's informative, if difficult-to-read-if-not-in-the-know chapter, "Wireless: the Next Generation." The ever-increasing complexity of telecommunications demands acronyms, if for no other reason than to keep sentences and paragraphs at reasonable length. Fortunately Brodsky spells out his acronyms on first use.

In his final chapter "Lessons in Creativity and Technology Development" Brodsky briefly describes eight lessons in science, business and invention that he thinks promote success and avoid pitfalls, including the time-proven chestnut: "Creative ideas threaten the status quo." Mac and Linux users will surely resonate with Lesson # 5: "The best technology is not always the most successful."

Brodsky's glossary does a superb job of providing non-technical definitions for technical terms. Though an extensive bibliography is included, the book's main text generally lacks footnotes which would tie information and ideas to particular publications. For example, Brodsky states that de Forest "amassed more than 300 patents." This figure is suspiciously round and even, and there is no reference to the source of the information. So this is not an academic history. Instead, think of reading it like spending an extended evening or weekend in pleasant conversation with a friend about the myriad aspects of old and new wireless.

How Silicon Valley Came To Be
Timothy J. Sturgeon. (Published as Chapter One in: *Understand-
ing Silicon Valley: Anatomy of an Entrepreneurial Region*,
Martin Kenney, ed. Stanford University Press, 285 pages, 2000).
http://ipc-lis.mit.edu/globalization/Silicon%20Valley.pdf
AWA Journal, v49 n3, July 2008, p. 24.

This site makes a nice complement to Morgan's book (see page 51). Though it covers much the same subject and time, where Morgan relies on description Sturgeon tends to analysis and interpretation, particularly of the military and corporate influences on western U.S. electronics development. He sees the trend toward cooperation among early Silicon Valley companies as a positive reaction to the domination of the electronics industry by "eastern" giant RCA. He sees company spin-off as a desirable, built-in feature of Silicon Valley companies that was not encouraged in the east. He also wants readers to know that Silicon Valley did not start with Fairchild Semiconductor or even with Hewlett and Packard, as many people believe. His historical starting point is Cyril F. Elwell's founding of Federal Telegraph Company in 1909.

Sturgeon's over-all conclusion is that the Silicon Valley success, unique in historical precedence and location, is unlikely to be duplicated elsewhere. This opinion has not precluded cities, univer-sities and even countries from studying the Silicon Valley phenomenon with a view toward lucrative imitation.

(See also Jane Morgan, *Electronics in the West*, p. 51).

Impoverished Radio Experimenter: Tricks, Tips, and Secrets To Help the Builder of Simple Radios and Electrical Gear Achieve High Performance at Minimal Cost!
Lindsay Publications, Inc.; *www.lindsaybks.com,* 48 pages per
volume.
AWA Journal, v48 n2, April 2007, p. 37.

How could I have been "in" electronics over half a century and not have heard about Lindsay Publications? Yesterday a box arrived on my front porch containing several of their publications. The box contained nothing else, so I'm going to assume someone at Lindsay wanted me to review the enclosed items.

I have the inevitable objection of a man my age to books like this–print too small, pages too small, pictures too small.

On the positive side, the publisher seems to be odd, shy, intriguing, and definitely sporting a sense of humor. This assessment remained unchanged after a visit to the internet address shown above. I still don't know if Lindsay is an author, publisher, jobber, wholesaler, distributor or, as I suspect, all that and more.

But to the *Impoverished* series. Go online to read all the subtitles and get a sense of the wide variety of material covered–trf regens, how to make your own coil winding machines, tube substitutions, building an impedance bridge, active long wire antennas, etc. There are even directions for making your own slow-motion dial drive. Nice clear photos and beautifully-drawn schematics are welcome enhancements. In short, enough to keep the curious neophyte who has no money busy for the rest of her life.

Lindsay also provides reprints of radio documents, of which they sent me two–Fessenden's *Wireless Telegraphy* from the 1908 Smithsonian Institution Report (catalog price $6.95) and the Institute of Radio Engineers 1945 *Radio Pioneers* 64-page pamphlet ($8.95). Both reprints of top quality. A sidebar to the Fessenden book ad asks "Who was Reginald Fessenden?" and answers its own question–"A genius ten times smarter than Tesla ever hoped to be." Well . . . .

Also included in my box of Lindsay goodies was a copy of their 80-page catalog of technical books, dated January 2007. The catalog's subtitle is "Exceptional technical books for experimenters, inventors, tinkerers, mad scientists, and 'Thomas-Edison-types' ". This is a treasure. Remember spending hours visually devouring the pages of the Allied Radio catalog in the 1950s? Now you can relive those golden days with the Lindsay catalog.

Lindsay's mantra must be something like: No matter what, Do It Yourself. Of course, there are various levels of involvement in DIY. For example, I enjoyed watching all the Menassa videos listed above (see page 7), even while realizing I will probably never get involved at the level he is. No, my mantra is much closer to the philosophy embodied in this often-reprinted couplet:

> *Pre-drilled chassis, my cup of tea,*
> *Heath kits surely were made for me.*

But *from scratch* DIYers *must* have a copy of Lindsay's catalog. Here they will find books that will guide them through electronic projects, electroplating, playing the violin, brick manufacturing, deep hole drilling, making Stirling engines, "war toys for boys", and numerous other endeavors, some of which are of such a nature that I won't list them here, knowing they wouldn't get past my editor.

I was tempted by many of the books, but will be ordering just one–Scherz's *Practical Electronics for Inventors*. The catalog warns– "Almost 1000 pages . . . . It's a huge book. A pain in the butt to pack." Stay tuned. I hope to review it in a future column. (See page 127).

Index to AWA Publications "2004-1/2 Issue"
Ludwell Sibley, The author, 90 pages, 2004.
Old Timer's Bulletin, v45 n4, October 2004, p. 60.

Only Lud would subtitle his monumental 8,000+ entry index with a "numerical acronym." These 90 pages at once make your *OTB* and *AWA Review* collections double their value by providing ready access to the information in them. Also included are entries for Tube Collectors Group *Fact Sheets*, the *Fifty Years of AWA* booklet and other AWA publications, January 1960 through 2004-1/2. If you know it's *somewhere* in one of these, you should be able to quickly find exact volume, issue and page numbers using this classified index. All AWA die-hards will want this one.

Index to Radio and Electronics Patents: A Keyword Index to the
Titles of Over 6400 U.S. and British Patents Issued to 100 Inventors,
With an Emphasis on Radio and Television
David W. Kraeuter, UMI, 512 pages, 1995.
Pittsburgh Oscillator, v10 n4, December 1995, p. 16.
(Yes, I had the cheek to review one of my own books).

Even with a computer doing much of the sorting and printing, it is
evident that a great deal of work went into this book, and it is an
obvious labor of love.

The long subtitle is an accurate description. The 512-page volume
can be used as an index to the three earlier bibliographies upon
which it is based or as a stand-alone index to 6400+ U.S. and British
patents dating from 1830 to 1980.

Patent titles have been entered in the original form and, in most
cases, in one or more inverted forms to provide access to internal
keywords. For example, Edwin Armstrong's 1920 patent for
"Receiving high frequency oscillations" is entered in the following
three forms:

High frequency oscillations, receiving. 1,342,885. 1920.
  Armstrong.
Oscillations, receiving high frequency. 1,342,885 1920.
  Armstrong.
Receiving high frequency oscillations. 1,342,885. 1920.
  Armstrong.

The index is not without limitations. The most important of these
is inherent in the general and often vague nature of patent titles. For
example, there are 12 patents listed for *acoustic device* and the only
way to determine the exact nature of these devices is to consult the
original patents. In some instances the situation is improved by the
addition of cross references, but there are not nearly enough of these.

Since an attempt was made to include all patents issued to every
inventor covered, some of the patents listed are for inventions
entirely outside radio. Most of the patents of H. P. Davis and H. P.

Maxim are examples. Also, the patents of Edison, Tesla, and Morse are absent, though the author does cite previously-published patent lists for Edison and Tesla.

Nevertheless, nothing has appeared yet that comes even close to what this volume does for radio patent literature, and the index will find many uses, such as verification of patent numbers, dates, and country of origin. Of course the book can also be used to gain quick entry into the extensive U.S. and British patent volumes.

The arrangement of the index permits one to obtain thumbnail chronologies of particular devices. For example, the first patent listed in the book specifically for *television* is British 269,834 to John L. Baird in 1927. By 1929 there were 20 patents listed for this device, then 14 in 1935, falling off to one in 1945. The earliest patent listed here for *color television* also went to Baird–British 473,303 issued in 1937.

The index is so extensive (over 10,000 entries) that it also invites random browsing. Oddities pop up unexpectedly–Gernsback's hydraulic fishery (2,718,083), de Forest's transistor (2,735,049), Alexanderson's computers (2,417,229; 2,984,414), Dubilier's mousetrap (1,630,241), Farnsworth's nuclear fusion apparatus (3,386,883), Sarnoff's secret signaling system (2,455,443), and Fessenden's television (2,059,221) are examples.

This reference book will find years of use by librarians, researchers, historians, and anyone with an interest in patent literature and the technical side of radio's history.

Invention That Changed the World: How a Small Group of Radar Pioneers Won the Second World War and Launched a Technical Revolution
Robert Buderi, Simon & Schuster, 575 pages, 1998.
OTB, v44 n3, August 2003, p. 60.

Many inventions changed the world, but Buderi argues convincingly here for radar being the decisive factor in WW II. He goes further and argues that radar's birth in the '30s and frantic development during the War were necessary prerequisites to what he calls "the dramatic, concentrated example of the endless intertwinement of science and technology." He links radar to such post-war advances as the transistor, the maser, nuclear magnetic resonance, tracking of commercial flights, television networks, radio astronomy, particle accelerators, etc. About half the book is devoted to these "children" of radar. (Remember when the microwave oven was called a radar range?)

Buderi always attempts to illuminate his subjects through biographical information gathered from interviews with many of the main characters. For example, during the War much of the United States' radar development work was done at MIT's Radiation Laboratory (so named as camouflage). Scientists there soon learned to have a reply ready when they took their ideas and suggestions to their supervisor, Isidor Rabi. Buderi tells us that Rabi inevitably wanted to know, "How many Germans will it kill?"

This book is a major popular–i.e., non-technical–history of its subject. It has already become a standard source in its field and is in well over a thousand libraries worldwide.

(See also Robert Watson-Watt's *Pulse of Radio* on page129).

J. B. Johnson and the 224A CRT
John B. Johnson, Tube Collectors Association, 32 pages, 2003.
OTB, v44 n3, August 2003, pp. 60-61.

Another pioneer has been rescued from semi-obscurity by this TCA publication, this time one of the inventors of the cathode ray tube. The editor states: "If one were to declare a 'Father of the Oscilloscope,' Johnson would be in the running at least equal to Allen Du Mont."

The 224A cathode ray tube was introduced in 1922 when Johnson was in his mid 30s and working at Western Electric. WE published a 41-page treatise, probably written by Johnson, giving detailed instructions for operating the 224A. A 24-page extract of that treatise makes up the bulk of Special Publication No. 5. The text and figures detail the introduction, description, installation and operation of the tube, and the pamphlet ends with a 75-item bibliography of publications (1897 to 1925) related to the cathode ray oscillograph.

The 224A was not Johnson's only contribution; he is probably best known for the concept of Johnson noise, a.k.a. thermal noise, the noise produced by the agitation of electrons in conductors and semiconductors.

Japanese Radar and Related Weapons of World War II
Yasuzo Nakagawa, Aegean Park Press, 103 pages, 1997.
AWA Journal, v46 n4, October 2005, p. 59.

A search of the online Library of Congress catalog for "United States" and "radar" returns 325 records. A similar search for "Japan" and "radar" returns only 19 records, and many of those are in Japanese. These results provide a good indication of the difficulty in finding information on the development of radar in Japan.

Nakagawa's book is a start in solving this problem. He shows us that the development of radar in Japan was every bit as hectic as in the U.S. and Europe and was fraught with many of the same problems, such as lack of basic science in the subject and the difficulty in getting military and government agencies to realize radar's potential.

Just as radar technology was secretly exchanged between England and the United States during the War, so too Germany undertook to get its radar expertise delivered to the Japanese. These submarine missions in 1941, '42 and '43 involved high adventure and loss of life and were not entirely successful. As a result, production of German radar (the Würzburg) was delayed in Japan until early 1945. Radar advancement in Japan was never as robust as in Germany or Allied countries, and Japan's most advanced radar systems "were not destined to see any important war service."

With author and subject indexes, a table listing Japanese radars, and 30 black-and-white photographs of Japanese radar equipment, installations and personnel.

John Logie Baird: A Life
Antony Kamm and Malcolm Baird, National Museums of Scot
    land Publishing, 465 pages, 2002.
OTB, v44 n2, May 2003, p. 59.

What do you think of first when you hear the name John Logie
Baird? Mechanical television. But the adjective should not be
allowed to overshadow the noun. Baird, as an inventor of
television–perhaps *the* inventor of television–may have started out
with primitive cardboard equipment. However, during his relatively
short lifetime he saw television through to electronic scanning on a
patented color CRT of his own invention. Baird's life was
characterized by genius, struggle, ill health, limited commercial
success and great creativity.

Now Baird's son and Antony Kamm have provided us with a
richly-detailed life of the inventor, ready to take its place beside
Russell Burn's biography (see next review). It is obvious from the
finished product that an enormous amount of work went into the
creation of this outstanding book. Two examples will suffice to show
the extent the authors went to in ascertaining the facts of Baird's life.
Baird's television career started in the town of Hastings in 1923. In
attempting to determine exactly where Baird lived at this time, the
authors consulted Baird's memoirs, city directories and even
electoral rolls. They conclude, nevertheless, that the exact location
could not be determined.

Baird's career (and life) ended in Bexhill in Sussex, not far from
Hastings. The authors tell us that when television finally came to
Bexhill years after Baird's death, reception was "consistently
reported to be bad." But Baird and Kamm are not content to let the
matter rest there; one of the book's 1100 endnotes informs us, with
dated documentation, that reception in that area today still requires
a booster.

Other niceties include a genealogical chart of the Baird family, an
appendix detailing Baird's involvement with the supernatural, and a
bibliographic essay comparing various copies of Baird's autobiography. There are 60 monochrome photographs. All are captioned

as meticulously as the book is written. My favorite is of Baird, indoors, ". . . in an overcoat to combat the chill of August in Berlin . . . ."

John Logie Baird obviously will be welcomed by Baird enthusiasts as well as television historians. But it undoubtedly will also be profitably used by anyone who enjoys reading biography or anyone who wants to see how good biography is done. Highly recommended.

John Logie Baird, Television Pioneer
Russell Burns, Institution of Electrical Engineers, 417 pages, 2000.
OTB, v42 n4, November 2001, p. 27.

Another thoroughly-researched, professional and beautifully-produced volume from the IEE. Burns knows his subject well and has contributed several other volumes to IEE's outstanding series. (See list below). Burns also had access to material not available to other Baird biographers, particularly relating to Baird's business dealings.

John Logie Baird considered himself to be the inventor of television. So do I. By 1932 he had devised and demonstrated wired and wireless transmission of moving images, noctovision (infra-red television), two-way television, phonovision (storage of television images on phonograph records), transatlantic television, color television, stereoscopic television, zone television (using separate channels to produce separate areas of the picture) and large-screen (cinema) television, all using scanning discs.

But Baird did not invent television out of thin ether, so to speak. He was preceded and influenced by the work of many investigators. Burns provides a list of these on page 36. Just as the list of inventors of the radio tube could be extended back to Thomas Edison in 1883, so too the list of television inventors could be extended back to Willoughby Smith, who in 1873 discovered that the resistance of selenium changes depending on how much light strikes it. Paul Nipkow designed but did not build a television system in 1884. Baird, however, brought everything together and patented and demonstrated working instruments.

Like many others, Baird had a few false starts early in his career. Having decided not to follow his father as a minister, he pursued a career as a purveyor of cheap soap. (Years later Baird titled his autobiography *Sermons, Soap and Television*.) In 1919 Baird traveled to the West Indies, partly to find weather beneficial to his health. There, in an open hut in the jungle, he tried his hand at manufacturing citrus and guava jams to be marketed locally and in England. An immediate problem arose: flying insects, attracted by the smell of the cooking jam, fell in it. Did the jam go to market anyhow?

Returning to England in 1920, Baird began to pursue his television ideas. Initially he had to make do with bare minimum resources, particularly money, equipment and working room. This forced him to invent with simple techniques and basic materials, and to exploit each material to its maximum use. This design practice carried over into his later work. For example, when it became desirable to combine sound and video reproduction in the same television receiver (instead of using a separate radio for the audio signal) Baird came up with the brilliant, though evidently unrealized, idea of having the scanning disc perform double duty as the speaker diaphragm! (See British patent 318,278).

In attempting to present an objective evaluation of Baird's work, Burns often compares Baird with other inventors, particularly Marconi, and with other companies. Here Baird, the small independent inventor, frequently outshone the work of the R&D departments of the BBC, RCA, EMI, etc.

Baird both benefitted from and suffered from being first with television. His many early successes blinded him and others to the fact that he came to television through the dazzling but dead end of mechanical generation and reproduction. In 1928 he wrote, "The use of the cathode ray is beset with the greatest difficulties, and so far, no practical success has been met with in its application." Thus he stuck by mechanical television too long after others realized its insurmountable limitation, the scanning disc.

Still, his head was as full of ideas about television as Edison's was about everything else, and in 1942 Baird demonstrated his three-gun color CRT. But from an historical perspective one wishes Baird could somehow have been "bumped over" into electronic television at a much earlier point in his career so that we could have reaped more of the benefits of his thinking in that area. But of course, though history can be and often is rewritten, it cannot be relived.

This is the second volume of the IEE History of Technology series to be reviewed here. A review of a third volume in the series follows this one (See *Life and Times of A. D. Blumlein*, below). Others will be reviewed in future issues. The entire series is:

1. *Measuring instruments–tools of knowledge*, P. H. Sydenham
2. *Early radio wave detectors*, V. J. Phillips
3. *A history of electric light and power*, B. Bowers
4. *The history of electric wires and cables*, R. M. Black
5. *An early history of electricity supply*, J. D. Poulter
6. *Technical history of the beginnings of radar*, S. S. Swords
7. *British television–the formative years*, R.W. Burns
8. *Hertz and the Maxwellians*, J. G. O'Hara and D. W. Pricha
9. *Vintage telephones of the world*, P. J. Povey and R. A. J. Earl
10. *The GEC Research Laboratories 1919-84*, R. J. Clayton and J. Algar
11. *Metres to microwaves*, E. B. Callick
12. *A history of the world semiconductor industry*, P. R. Morris
13. *Wireless: the crucial decade, 1924-1934*, G. Bussey
14. *A scientist's war–diary of Sir Clifford Paterson*, R. Clayton and J. Algar (Editors)
15. *Electrical technology in mining: the dawn of a new age*, A. V. Jones and R. Tarkenter
16. *"Curiosity perfectly satisfied:" Faraday's travels in Europe 1813-1815*, B. Bowers and L. Symons (Editors)
17. *Michael Faraday's "Chemical notes, hints, suggestions and objects of pursuit" of 1822*, R. D. Tweney and D. Gooding (Editors)
18. *Lord Kelvin: his influence on electrical measurements and units*, P. Tunbridge
19. *History of international broadcasting*, J. Wood
20. *The very early history of radio, from Faraday to Marconi*, G. Garratt
21. *Exhibiting electricity*, K. G. Beauchamp
22. *Television: an international history of the formative years*, R. W. Burns
23. *History of international broadcasting*, Volume 2, J. Wood
24. *The life and times of A. D. Blumlein*, R. W. Burns

25. *History of electric light and power*, 2nd edition, B. Bowers
26. *A History of telegraphy*, K. G. Beauchamp
27. *Restoring Baird's image*, D. F. McLean

(Series editors: Dr. B. Bowers, Dr. C. Hempstead).

Legacies of Edwin Howard Armstrong
John W. Morrisey, ed. Radio Club of America, 321 pages, 1991.
AWA Journal, v48 n4, October 2007, p. 59.

The great value of this book is that it reprints 14 papers, most of them technical, written by Armstrong himself. His 1915 paper minutely detailing the audion's behavior is here, as is his scathing 1948 dissection of Stuart Seeley's ratio detector. Some of these reprints are of papers Armstrong presented before the Radio Club of America. Copies of their *Proceedings* are rare; even the great Linda Hall technology library in Kansas City, Missouri lacks them before 1950. Altogether the reprints included here represent a good chunk of the 26 papers by Armstrong listed in the book as a reprint of the bibliography that originally appeared in Lawrence Lessing's *Man of High Fidelity*.

Legacies, and Armstrong's reprinted papers, are organized according to his major inventions–the regen, superhet, and super-regen circuits, and, of course, wide-band FM. Each of these sections reprints at least three of Armstrong's papers, except for the super-regen section, which contains only one reprint.

A disappointment of the book is the poor reproduction of photographs from some of Armstrong's original articles. These are in some instances so small or so poorly presented as to be useless.

Also included are appreciations of Armstrong by many who knew him personally or studied his work. It would have been easy for their contributions to dissolve into hagiography, but a balance is provided by the translation of Lucien Levy's plea to be remembered for his own superhet work, and the inclusion of William Denk's "Two Rich Minds–One Poor Invention," an analysis of Armstrong's and Michael Pupin's U.S. patent 1,336,378.

The book closes with no fewer than 19 appendixes. One of these is a list of Armstrong's U.S. patents. The list unfortunately omits 17 patents, probably due to an editing or printing error. A correction to the list is being published in my book, *Ten Patents from Radio History*.

Life and Times of A. D. Blumlein
Russell Burns, Institution of Electrical Engineers, 534 pages,
  2000.
OTB, v42 n4, November 2001, pp. 28-9.

Burns has obviously written *the* Blumlein biography here, but read
the title carefully. The book actually has two subjects: Blumlein and
the times in which he lived. Thus a considerable portion of the book
deals not with Blumlein directly but with the electronic, corporate or
military worlds in which he worked. This is particularly true of the
second half of this long book, which could almost have been
published as a separate book on radar in World War II.

Some readers may not recognize Blumlein's name, though
probably all readers have benefitted, directly or indirectly, from his
many electronic inventions, innovations and developments. These
include the +45/-45 degree recording groove in stereo phonograph
records (what would Blumlein think of the CD?), the cathode
follower circuit (now called the emitter follower) and the slotted
television transmission antenna. Blumlein also invented the circuit
which he called the long-tailed pair, thus proving he had a sense of
humor.

Burns provides documentation to show that Blumlein's 1932/33
work in negative feedback "considerably predates" that of Harold S.
Black. (Nevertheless, Black's epiphanic realization of negative
feedback occurred in New York City in 1927). Blumlein's 120+
patents were for telephone, monophonic and stereo recording,
measurements, antennas and cables, power supplies, cathode ray
tubes and circuits, DC restoration, modulation, AGC, delay lines and
radar.

The publisher's blurb on the back cover of the book refers to the
biography as "meticulous" and there is no arguing this point. For
example, in the chapter on Blumlein's personality, Burns discusses
Blumlein's sense of humor. But first he takes a page or so to review
classical theories of humor, quoting Hobbes, Schopenhauer, Pascal
and Sir Philip Sydney!

Blumlein lived in the Park Royal section of London during the blitz of WW II. Always the innovator, he redesigned the window attachments in a room of his house according to his ideas about safety during bomb blasts. When in fact a bomb buried itself in the back garden a few yards from the house before detonating, the blast suction pulled the windows out of the room as Blumlein planned. The occupants of the room, including Blumlein's wife, were unhurt but described feeling the air being sucked from the room. Blumlein also installed a warning system in the attic of his house to detect the presence of incendiary bombs that might have penetrated the roof of the building.

One of the reasons for Blumlein's relative obscurity among radio enthusiasts may be the brevity of his career. Born in 1903, he died in 1942 in a military airplane crash. The purpose of the flight was to test the $H_2 S$ radar system that Blumlein and others had designed. His death was called a "national disaster" by the British Secretary of State for Air.

A classic danger in writing biography is the over-valuing of the subject's contributions and historical stature. Indeed Burns refers to Blumlein with the phrase "the greatest British electronics engineer of the twentieth century." This book goes a long way toward validating that claim. Its very thorough treatment (some will say too thorough) assures that it will remain *the* Blumlein biography for years.

Lightning Man: The Accursed Life of Samuel F. B. Morse
Kenneth Silverman, Alfred A. Knopf, 503 pages, 2003.
OTB, v45 n4, October 2004, p. 60.

Even though Morse's simple device of wires, switches and electromagnets used low-voltage batteries[1], he became known as "Lightning Man" because of the conflation in the public's mind of equally-mystifying lightning and electricity.

Silverman does not explain the phrase "accursed life" from the subtitle of his book. He has no need to. Morse struggled his entire adult life trying to establish and maintain his reputation–first as a portrait painter and then as an inventor. Like many successful inventors before and since, Morse became a nexus of insight, ambition, wealth and ego. He was not entirely likeable. Driven (and tainted) by his early strict religious training, he came to hate Catholics and see Catholicism as an attempt to undermine the American government. He thought he found Biblical evidence that condoned slavery and published tracts in support of it. He neglected his children for his work and sometimes failed to give his associates in invention proper acknowledgment for their contributions.

Morse's life and struggles remind us that invention and people's reaction to it could be just as unsettling and rancorous in the 19th century as it was in the 20th. Feelings for and against Morse ran high, even after his death in 1872. A life-size statue of him remains today in New York City's Central Park. But Amanda, the widow of Morse's closest colleague in invention, Alfred Vail, spent most of her adult life trying to gain recognition for Vail's telegraph work and weaken Morse's.

Did Morse invent the telegraph? You might as well ask if Marconi invented the radio[2]. But, as Silverman shows, Morse, like Marconi, got his system together in working order, patented it, made it available in usable (i.e., commercial) form, and continually demonstrated its value. Morse's system was also able from the beginning to record transmitted information, unlike the needle telegraph of Cooke and Wheatstone, which Morse referred to as a "semaphore"[3].

The world is indebted to writers like Silverman who are willing to devote years of research, thought and writing to the production of biographies such as this thoroughly readable one.

1. Morse telegraph systems operated with about 48 volts. See S. F. B. Morse, *Paris Universal Exposition, 1867. Reports of the United States Commissioners. Examination of the Telegraphic Apparatus and the Processes in Telegraphy*, Washington, DC: Philp & Solomons, 1869, p. 71.

2. For more on Morse's telegraph predecessors, see page 72-73 above.

3. Morse, *op. cit.*, p. 147.

Love Letters to Spike: A Telegrapher's Lament, With a Brief, Eclectic History of Communications in the Seacoast
Bill Holly, Placenames Press, 84 pages, 2004.
AWA Journal, v47 n3, July 2006, p. 41.

Ah, yes, to be simultaneously homesick and lovesick. Such was the plight of Herbert D. Waldron at the Portsmouth, New Hampshire Navy Yard telegraph office in 1914. His homespun, ingenuous letters sent back to his beloved "Spike" (a.k.a. "Gracie Darling") in Hartford, Connecticut, had they been made public at the time, could easily have taken the romance out of any young person's notion of becoming a telegrapher in a faraway land.

The prosaic content of the letters is given much life by the research and commentary provided by Holly on the contemporary state of telegraphy on the New England seacoast.

With about two dozen black-and-white photographs.

Lulu Website: www.lulu.com
AWA Journal, v48 n4, October 2007, p. 58.

Everyone, it's been said, has a book in him or her just waiting to be written. If that's true then the AWA membership collectively represents a radio history library of several thousand volumes, since each AWA member has had unique experience with radio.

So where's your book? And don't say you haven't written it because you know how difficult it is to find–and please–a publisher. That onerous burden has been eliminated for all would-be authors by Lulu, the free on-line publisher. Simply put–you write it, they'll publish it. And they'll pay royalties. In fact, you the author determine the royalty amount. (Slight pause here while AWA members rush to their word processors).

As soon as you finish your book (considerably longer pause while AWA members finish their books) all you have to do is log on to the internet and send Lulu an electronic copy of the book. They'll want to know certain particulars, such as page size, hardcover or soft, etc. They want to receive your book in Portable Document Format, a.k.a. PDF. Those of you who wrote your books using WordPerfect X3–no problem. Just click on "File", then "Publish To", then "PDF". Or you can ask Lulu to convert the file for you.

Lulu strongly suggests that at first you order only a single copy of your book, for proofreading purposes. That's because it's easier to see errors in paper copies than in on-screen copies. After you proof your paper copy, make any corrections necessary to your PDF file, send it back to Lulu, and then sit back and let the royalties roll in. Don't need the royalties? You can specify that and Lulu will then sell your book for printing costs only. A 60-page paperback book with 8 ½ x11 inch pages, including photographs, etc. will sell for less than $6 per copy. Not feeling quite so magnanimous? Set your per-copy royalty at whatever suits you. But be warned–if you set it at $100 per copy, sales might atrophy.

Tempus fugit. Still, if you send Lulu your PDF file today, I just might be able to review your book in the next issue of this column.

Man Who Changed Everything: The Life of James Clerk Maxwell
Basil Mahon, Wiley, 226 pages, 2003.
AWA Journal, v46 n2, April 2005, pp. 57-58.

"One scientific epoch ended and another began with James Clerk Maxwell." Thus spoke Albert Einstein and this short, very readable biography shows why. Maxwell was born in 1831 and died of intestinal cancer in 1879. In between he managed to . . . well . . . change everything. As a boy of three he demanded to know how things worked, pestering his parents with his standard question, "What's the go o' that?" (Isn't it nice that Maxwell's exact words as a three-year-old boy appear in the *AWA Journal* 170 years later?) By his early 30s he knew how things worked and propounded his theory of electromagnetism. His four formulas (tamed by Oliver Heaviside) are still in use by electrical engineers.

The seminal theoretical truth that Maxwell laid down in 1864, however, was not completely accepted until years after his death. Indeed, Mahon points out that even the great Lord Kelvin, who lived until 1907, probably never did understand Maxwell's electromagnetic theory. But in Germany Hermann Helmholtz encouraged his student Heinrich Hertz to investigate further. Hertz was so successful that "Eight years after his [Maxwell's] death, Maxwell's electromagnetic theory had been emphatically verified."

Maxwell enthusiasts (if not outright worshipers) are now known as Maxwellians[1]. They may be saddened to learn that he was not perfect. Mahon reports that Maxwell was notably lacking in some lecture skills and was known for making relatively simple errors in his calculations. Nevertheless his thinking was profound, and we can get some idea of Maxwell's posthumous scientific stature from this part of Heaviside's famous assessment of him: "His soul will live and grow for long to come, and hundreds of years hence will shine as one of the bright stars of the past, whose light takes ages to reach us."

1. For more on Maxwell's belated rise into scientific super-stardom, see Bruce J. Hunt's *The Maxwellians*. It shows the work of four

famous Maxwellians–FitzGerald, Heaviside, Hertz and Lodge–and is reviewed here on page 110.

Man Who Invented the Twentieth Century: Nikola Tesla, Forgotten Genius of Electricity
Robert Lomas, Headline Book Publishing Company, 248 pages, 1999.
OTB, v43 n2, May 2002, p. 52.

Lomas's book is a short, popular account with no footnotes. Seifer's scholarly study, *Wizard,* is based on his Ph.D. dissertation and is heavily documented (see page 199). Both books present a life that can only be described as fascinating.

Poor Tesla. Nothing in his life was ordinary. He was by turns supremely brilliant and naive, wealthy and destitute, legally and illegally robbed, and a visionary genius who sometimes earned his keep by digging ditches. He never married and apparently formed few close relationships. He lived in hotels for most of his adult life, and often shared his rooms with pigeons. He learned through an article on the front page of the November 6, 1915 *New York Times* that he was to share a Nobel Prize with Thomas Edison. In a subsequent interview he expressed his gratitude to the world for finally recognizing the value of his work, only to learn later that the *Times* story was false.

Tesla (pronounced TESH la) gets an A+ in electricity, but a much lower grade in business acumen. Legend says he sold the 40 patents for his AC polyphase system to Westinghouse for $1,000,000 (some writers report a much lower figure), plus a royalty of "one dollar per horse-power." Later he either relinquished this royalty, surely worth millions, or sold it back to Westinghouse for a paltry sum. Still later he testified that he thought at the time that Westinghouse was to have the royalty back only temporarily. For much more detail on these shenanigans, see the Seifer book.

One of the reasons Tesla's life is so intriguing is that so much of it, including the exact date of his death, ends in mystery. Some of this mystery is attributable to what Seifer says was Tesla's "irritating habit of ... seeing projects complete before they actually materialized ...."

Surely Tesla's most important idea, i.e., the idea that had the most potential for benefitting humanity, was the transmission of electrical power without wires. Tesla sold his patents for this system to J. Pierpont Morgan. Morgan "sat" on these patents and thus stymied further development of them, possibly to protect his own considerable financial investment in Tesla's earlier wired system of AC power transmission.

So here is the question: could Tesla's system of wireless power transmission be made practical? If so, why has no one done it? Patents for the system were published 100 years ago. The fact that no one has achieved widespread wireless transmission of power to date would seem to indicate that Tesla's scheme cannot work.

Readers may argue forever over Lomas's list of major inventions claimed for Tesla: radio, the hydroelectric generator, AC power, fluorescent light, the rotary engine, the bladeless turbine and radio imaging. But who could quibble with an author who dutifully reports one of Tesla's lesser-known inventions, a significant enhancer of peristalsis? Tesla used the device, a variable-frequency vibrating platform upon which one stood, to successfully treat his indisposed friend Mark Twain.

These books can't help but make many of their readers Tesla simpaticos.

Marconi's Atlantic Leap
Gordon Bussey, Marconi Communications, 96 pages, 2000.
OTB, v42 n2, May 2001, p. 63.

Do we need a whole book about sending the letter 'S' across the Atlantic Ocean? In this case it is justified for no other reason than the singularity of the event. In 1895 Marconi was sending radio waves over a distance of a few yards; on December 12, 1901 he sent them from Europe to America, thus proving that radio waves could follow the curvature of the earth.

One might think that Marconi's accomplishment would have been welcomed immediately by humanity, and he did receive recognition from nations worldwide. But on the very day he publicly announced his success he also received a letter from the Anglo-American Telegraph Company of Newfoundland informing him of their monopoly on telegraphic communication in that colony and calling upon him to discontinue his work and dismantle his station. Marconi immediately suspended his experiments. Alexander Graham Bell would soon offer Marconi the use of his estate in Nova Scotia, where Anglo-American had no jurisdiction.

Bussey's book is greatly enhanced by the inclusion of 70 reproductions of photographs. About half these are contemporary pictures of the transmitting and receiving installations at various stages of completion. One could describe a rigger working at the top of a 200-foot aerial mast, but it is much more meaningful to actually picture him. Other photographs reproduce manuscript documents, printed records, maps, etc. The text and pictures of this little book combine nicely to give the reader a most rewarding "I was there" feeling.[1]

Some people will come to this book wanting to determine what frequency Marconi used for his historic transmission. Bussey quotes Marconi as saying (in 1908) that it was about 366 metres (820 kHz), but Bussey reminds his readers that "no convenient method of measuring station wavelengths had been evolved at that time . . . ."

Twenty-first century readers may also have to be reminded of other facts about Marconi's achievement. For example, the famous 'S' went only one way across the Atlantic. Marconi had no high-power transmitter in Signal Hill, Newfoundland. Also, the receiving equipment at Signal Hill was passive; it had no ability to amplify. It could be thought of much as we think of a crystal set today, except in place of the crystal was a none-too-reliable coherer. Marconi increased the sensitivity of his receiver by burying metal plates in the earth for a ground, and using a large kite whose 500-foot tether acted as an antenna.

And here is a question that Bussey's book did not (and could not) answer: how did Marconi feel when he first heard those few faint signals from far across the sea?

1. For more information on Marconi's transmitter, see Desmond Thackeray's article "First High-Power Transmitter at Poldhu," *AWA Review*, volume 7, 1992, pp. 29-45.

Maxwellians
Bruce J. Hunt, Cornell University Press, 266 pages, 1991.
AWA Journal, v50 n2, April 2009, p. 49.

Maxwell's treatise propounding his theory of electromagnetism was published in 1873. At his death in 1879 Maxwell was beginning his revision of the work. As he left it, his theory was profound, abstruse and, in some instances, incorrect. Hunt refers to it in his foreword as "rambling and difficult." As a result the theory was not immediately accepted by the world of science.

The Maxwellians were those physicists–Oliver Lodge, G. F. FitzGerald, Oliver Heaviside and Heinrich Hertz–who continued Maxwell's work after his death. All four men admired or even idolized Maxwell. They refined, enhanced and corrected his work to such an extent that the curmudgeon Heaviside was able to say, tongue in cheek, that Maxwell himself was by then "only ½ a Maxwellian." Lodge built physical models to represent Maxwell's theory and explicated his work in popular publications. FitzGerald, the theoretician, was also the group's facilitator who made suggestions that stimulated the others. Heaviside played his part as the confident, self-taught mathematician. Hertz, of course, vindicated the work of the other three by demonstrating electromagnetic waves in 1887-88. By the end of the century Maxwell's theory had become thoroughly accepted.

Two Irish mathematicians, James MacCullagh and Joseph Larmor book-ended the Maxwellians, with MacCullagh's work in the 1830s greatly influencing FitzGerald, and Larmor proposing the existence of the electron in the 1890s. In between William H. Preece of the Post Office Telegraph Department played the part of the villain by attempting, with considerable initial success, to suppress the eccentric Heaviside's writings. Heaviside ultimately appealed to Lord Kelvin to resolve the matter. Heaviside also reduced Maxwell's complicated formulas for electromagnetism to the essential four that are used today.

Hunt tells the story of Maxwell's posthumous, belated rise to scientific superstardom knowledgeably and clearly while–nice for some of us–generally avoiding mathematics. With this cogent and lucid book, Hunt in fact does for the Maxwellians what the Maxwellians did for Maxwell.

McCandless and the Audion
Gerald F. J. Tyne, Tube Collectors Association, 12 pages, 2003.
OTB, v45 n1, February 2004, p. 45.

Who gets your vote as inventor of the radio tube–Edison? Fleming?
De Forest? Before you answer you might spend some time with this
latest publication/audio CD from TCA. H. W. McCandless (b. 1866)
did not invent the radio tube, but he did craft to order many of the
audions that were used by de Forest, Gernsback, Marconi,
Armstrong, etc.

Too bad he didn't etch HWM on the envelopes of the audions he
made. But that type of first-hand brush with history is provided here
amply in the audio recording in which McCandless reminisces about
his relationships with some of the biggies in early radio. The "live"
audio comes complete with missed takes and long pauses by
McCandless as he struggles to recall his long-ago pioneering days.
As time passes such productions as this one can only gain in value.

Mouser Electronics: a tti company. February-April 2007 Catalog 629. "The Newest Products For Your Newest Designs." *www.mouser.com*, 1,840 pages, 2007.
AWA Journal, v48 n4, October 2007, p. 59.

I suppose one could review this electronic parts supplier's catalog as a book, but I want to review it as a human artifact. Looked at like that it's a pretty amazing one. This dense object weighs in at three pounds, eight ounces. There are 1,840 Bible-thin-pages of specifications for electronic parts, each page containing dozens or hundreds of facts, including illustrations. We elderly-or-nearly-so appreciate the use of a magnifying glass in viewing them. Look, for example, at the BOMAR Between Series Adapters on page 953–there are 45 postage-stamp-size line drawings of connectors, each ever so slightly different from the next. The back of an earlier catalog boasts "Nearly Every Connector Under the Sun" and it would be hard to dispute this claim (but I still can't find a male TNC to female adapter for the VHF whip antenna on my Drake R8B).

Who made up this artifact, and what kind of computer and program did they use to produce it? I'd like to meet the catalog production staff at Mouser just to verify that they are real people. And I'd like to ask them a few questions, like do they relax in a rock band after work and, if so, are they available for church potluck suppers on Wednesday evenings? Are the people who proof-read the catalog still sane when they get to page 1,839? Who has to take the catalogs to the post office?

Anyhow, I decided to put Mouser to the test. Using Catalog 629 I found the Mouser stock number for some batteries and other minor parts I needed and logged on to their internet site. I typed in the part numbers and was quickly taken to a description of the parts. My order was completed rapidly and when the box arrived on my front porch a few days later it contained exactly the parts I had ordered. That fact alone was impressive.

Mouser is also to be commended in this marketing day and age for still having no minimum order requirement, and that policy should be the delight of every bored and near-broke schoolboy who wants to beef up the bass on his boombox, build a better bomb, or just buy a binding post. So one could, I suppose, buy a single (bulk packaging) 1/4-watt one-ohm resistor for 5¢ (plus postage). If you want to simultaneously cut unit costs and stock up, order 25,000 of them (tape and reel packaging) and you'll get the unit price down to just half a penny (plus postage).

Don't look in the catalog for 01A's, or even 35W4's. There are no tubes to be had. But if you need a standard 1N914 switching diode in a DO-35 package, you can get one for 6¢. A diode for 6¢–what would Edison or Fleming have thought?

National Museum of Broadcasting
http://www.nmbpgh.org
AWA Journal, v51 n1, January 2010, p. 59.

Never heard of the National Museum of Broadcasting? That's probably because it doesn't exist–yet. But if some radio history enthusiasts have their way, that museum will become a reality somewhere in southwestern Pennsylvania.

Pittsburgh is rife with historical resources appropriate to a broadcasting museum. In the 1890s Reginald Fessenden, inventor of the heterodyne method of broadcasting now used universally, worked at what is now called the University of Pittsburgh. Vladimir Zworykin invented his electronic television system in Pittsburgh. Pioneer radio station KDKA, known throughout the world, is still in operation in Pittsburgh. World-wide short wave broadcasting was developed and tested in Pittsburgh in the 1920s. Pioneer public educational television station WQED, with Fred Rogers's long-running program *Mr. Rogers's Neighborhood*, was founded in Pittsburgh.

And then there was Frank Conrad, who worked as a radio engineer with the Westinghouse Company. His experimental broadcasts from his garage in 1919 and earlier showed Westinghouse that there was a market for radio broadcasting. Soon KDKA would make its famous first broadcast on November 2, 1920.

Conrad died in 1941; sixty years later his historical garage was taken apart brick by brick and put in storage. A major project of the National Museum of Broadcasting is the reconstruction of the garage, preferably at a site also appropriate for the establishment of a broadcasting museum.

The Pittsburgh radio history enthusiasts, whose goal is to preserve the birthplace of the broadcasting industry, have a lot of motivation. They have the garage waiting in storage. They've investigated several prospective sites for the rebuilding of the garage. They are now lacking only one thing to be successful.

Go to the webpage listed above to find much more information about the proposed museum and the Conrad Project.

9XM Talking: WHA Radio and the Wisconsin Idea
Randall Davidson, University of Wisconsin Press, 405 pages,
  2007.
AWA Journal, v48 n3, July 2007, pp. 30, 37.

The publication of this book reminds us of two things we have to
be thankful for: first, that Davidson was willing to invest the
considerable time and effort required to write the book and, second,
for the "Wisconsin idea" itself–the idea that right from the start
WHA would bring "the educational riches of the university to all the
state's residents."

WHA began as 9XM, broadcasting regularly from the University
of Wisconsin's physics department in 1921. The universality of the
Wisconsin idea resulted in WHA broadcasting such things as
directions for building a radio, of course, but also, for example, it
profiled and transmitted from a sorghum mill and offered a
typewriting course over the air. WHA's mission was enhanced in the
1930s through the establishment of the Wisconsin College of the
Air–"At their radios, note-books in hand, storing up rich wisdom."
The sign outside WHA's transmitter in 1932 proclaims in large
capital letters–VISITORS WELCOME. This is another indication of
public radio at its best.

KDKA Pittsburgh, KCBS San Francisco and WWJ Detroit fans
take note: WHA also has claims on the "Oldest/First Broadcaster"
title. Davidson presents his case for 9XM-WHA in a postscript to the
book, and references the 1977 Baudino/Kittross article in the *Journal
of Broadcasting.*[1]

9XM Talking contains about 70 photographs, three appendixes,
source notes and bibliography, and a general and station call letter
index. Its form and content could well serve as a model for future
radio station histories.

1. Joseph E. Baudino and John M. Kittross, "Broadcasting's
Oldest Stations: An Examination of Four Claimants," *Journal of
Broadcasting*, 21:1, Winter 1977, pp. 61-83.

Old Timer's Bulletin
CD Edition. Antique Wireless Association, 1997?
OTB, v42 n2, May 2001, p. 61.

Computer system requirements are: 486 or Pentium processor, Microsoft Windows 95 or 98, or Windows NT 4.0 with Service Pack 3 or later, 8MB of RAM on Windows 95 and Windows 98 (16MB recommended), and 10MB of available hard disk space. Your computer must also have Acrobat 4 on it. Don't worry if you don't have that–each disk comes with a free copy. (Mac users must visit *www.adobe.com* to download a suitable version). The discs are easy to load and use, for those of us willing to read instructions.

Among the first images you will see on each disc are a machine searchable index and a coded table of issue numbers which links to individual issues. I clicked on 011 (for volume one, number one) and up popped the first page of the first issue of volume one, containing Bruce Kelley's clarion call for wireless history enthusiasts to unite, complete with caps key stuck 'on', evidently intentionally. WELL FELLOWS, THERE APPEARS TO BE ENOUGH INTEREST TO TYPE THIS SHEET–SO HERE GOES. (What we would today call sexism was quite acceptable in 1960). Thus was launched the special-interest magazine that today has a readership of 4,000.

Scroll down to page two, if you like. In fact, if you want to, navigate all the way down to page 64 of the fourth issue of volume 37. Page after page of clear readable text and graphics–dozens of issues, thousands of pages, countless topics.

It's fun to see how quickly improvements were made on an issue-by-issue basis (this phenomenon is characteristic of grass roots magazine start-ups). By issue number two the caps key had been unstuck. By issue number four *The OTB* was ready for its first photographs, and these remind us of the great advantage of the Portable Document File (PDF) format used in the production of these compact discs–it permits reproduction of all types of material, including text in various fonts and typefaces, photographs, line drawings, etc.–literally a picture of each page. It's nice to see some features that remain constant. For example, the antenna towers and wires at the top of the first issue are pretty much the same as those reproduced in the issue you are holding now.

Too bad that the covers that were originally in color were reproduced on disc in black-and-white, presumably to save disc space. (In my opinion the Boucher November 1993 front cover rates as a classic–see below). Another disappointment to some will be the lack of search capability at the word level. Although Ludwell Sibley's detailed index is keyword searchable, the *OTB* text itself is not. As Marc Ellis, who produced both discs says, scanning and proofreading the entire text of the *OTB* in machine-readable format would have been "beyond the realm of practicality."

The discs are $49.95 each or $89.95 for the two-disc set. Too expensive? Suppose at your next flea market you have the most-

u n l i k e l y experience of finding a com-plete paper set of *OTBs* from the first issue through volume 37–no missing or loose pages, no missing issues, no graffiti, dog-ears, coffee stains, hand-written notes, torn-out ads, etc. What would you be willing to pay for them?

Oliver Heaviside: The Life, Work and Times of an Electrical Genius of the Victorian Age
Paul J. Nahin, Johns Hopkins University Press, 318 pages, 2002.
OTB, v45 n3, July 2004, pp. 13-14.

I would love to have known Heaviside. He was an electrical and mathematical genius who at his death left five volumes of published writings which stand today as monuments in the histories of both mathematics and electromagnetism. He taught himself mathematics and electricity so thoroughly that people around the world wrote to him for advice, so much so that he became known in the scientific world as the "Inexhaustible Cavity." On the other hand, Norbert Wiener once referred to him as "an undersized, hungry, deaf, cantankerous little electrician."

To be sure, the genius was also human. He quit school for good at age 16 and at 24 quit the only job he ever held and went home to live with his parents. When they died he lived alone, with less-than-adequate financial resources. He rarely left his house or neighborhood, but did enjoy bicycling. Possibly his only ride in a motor vehicle was the one which took him to a nursing home a few weeks before his death in 1925. And, if one of his last neighbors is to be believed, he was by then comfortable dressing in pale pink silk kimonos and painting his fingernails "a glistening cherry red."

Everyone knows Heaviside predicted what came to be called the Kennelly-Heaviside layer, known now as the E layer. But few know that Heaviside was also the inventor of the now-ubiquitous coax cable. This invention, the only one he actually bothered to patent (1880 British number 1,407) earned him the princely (and badly needed) sum of 100 pounds.

Another invention Heaviside should have patented and didn't was the long-distance telephone line loading coil. Money for that invention, about half a million dollars, went instead to Edwin Armstrong's mentor Michael Pupin, who did patent the device. Heaviside shared with Tesla an acute lack of business acumen. Or perhaps he simply didn't care about money.

He sprinkled his highly-technical writing with side comments on social or professional matters. Here he speculates on what today would be called psychophysics: "The fact that the brain is subject to material change and replacement during life does not debar the theory of partial dependence upon the inner world of the atom."[1] And on life in general: "Life is an essential property of matter. All matter is alive, even the deadest. All phenomena are natural phenomena."[1] Compare with Antony Flew's later, more cryptic, ". . . Stuff is all there is; while everything which is not stuff is nonsense."[2]

This is the paperback version of Nahin's 1988 book *Oliver Heaviside: Sage in Solitude*, with a new preface by the author. Heaviside's seemingly-pedestrian life was a fascinating one, and Nahin tells it very well and very knowledgeably. I can't imagine anyone starting to read this book and not finishing it.

1. Both quotes from Rollo Appleyard, *Pioneers of Electrical Communication,* Macmillan, 1930, page 255.
2. James A. Haught, *2,000 Years of Disbelief,* Prometheus Books, 1996, page 302.

Oliver Lodge and the Invention of Radio
Peter Rowlands and J. Patrick Wilson, eds. PD Publications, 241
    pages, 1994.
AWA Journal, v50 n4, October 2009, p. 28.

This book is somewhat rare in the U.S. *Amazon.com* doesn't provide it and "Find in a Library" (see page 25 of the April 2006 *AWA Journal*) lists a mere 17 U.S. libraries holding the book. I was unable to find any current address or telephone information for the British publisher. The verso of the title page lists PD Publications' address as 4 Ascot Park, Liverpool L23 2XH. But the book is well worth pursuing. Nine writers (including two of Lodge's grand-children) have contributed 16 essays and other items that illuminate Lodge's personal and professional life.

Why is the name Marconi, rather than Oliver Lodge, almost always thought of when one mentions the invention of radio? This book can be interpreted as an extended attempt to answer that question. In Oxford, England on August 14, 1894 Lodge lectured on and clearly demonstrated before the British Association meeting the transmission of intelligence via radio waves through space. At that time, Marconi was still struggling to transmit at his family's estate in Italy. Lodge's basic radio patent, issued early in 1897, recognized the need for tuning. Marconi's first patent, issued later, did not. Marconi bought the rights to Lodge's tuning patent in 1911 for £18,000[1].

In his extensive essay, "Radio Begins in 1894," Rowlands takes pains to present his case for the primacy of Lodge over Marconi as radio's true inventor. This is not hero worship; Rowlands backs up each of his opinions with reasons and many citations to the scientific literature. In fact, reading this book made me think that one would have to go to primary sources to get an account of Lodge and his 1894 radio work that was any more thorough than this one.

Lodge may have missed out on the radio invention prize because, unlike Marconi, he was too busy with his other work in physics, such as proving–or disproving–the existence of aether, or his work as a university administrator. Marconi concentrated single-mindedly on radio. In his dealings with other scientists and technologists, Lodge always worked under the unspoken code of the "Victorian gentleman," but Marconi had no such constraints on his behavior. Being a "pure" scientist, Lodge may also have simply failed to recognize the commercial potential of radio waves. Or, as is suggested in the book, Lodge's reputation in science may have been tarnished by his interest in spirituality, in which he investigated mediums and attempted to communicate with the dead. In this he conducted some experiments that would today be considered bizarre. Lodge lived from 1851 to 1940. Nevertheless, the chronology of his life that the book ends with coyly–but correctly–lists the conclusion of his last experiment as happening on May 19, 1954. And if *that* sentence appears bizarre to you, see pages 153 to 155 below.

1. G. R. M. Garratt, *Early History of Radio*, Institute of Electrical Engineers, 1994, page 69.

On the Short Waves, 1923–1945: Broadcast Listening in the Pioneer Days of Radio
Jerome S. Berg. McFarland & Company, Inc., 280 pages, 2002. (See page 57 of the August 1999 *OTB* for a review of this volume.)

Broadcasting on the Short Waves, 1945 to Today
Jerome S. Berg. McFarland, 496 pages, 2008.

Listening on the Short Waves, 1945 to Today
Jerome S. Berg. McFarland, 423 pages, 2008.

AWA Journal, v50 n2, April 2009, pp. 48-49.

We used to call Milton Berle "Mr. Television." With the completion of this trilogy on shortwave radio I am tempted to call Jerome Berg "Mr. Short Wave." Perhaps it was his training and experience in law that makes Berg treat his subject with the competence and thoroughness that is evident in these volumes. (You know you're dealing with a savvy author when, in the preface, he apologizes to James Watt for using the abbreviation *kw* instead of *kW*.)

In *Broadcasting*, Berg provides a year-by-year history of shortwave transmission since the War, highlighting hundreds of shortwave stations worldwide. But first he provides an overview of shortwave broadcasting, patiently and clearly explaining the differences among such esoterica as surrogate and nonsurrogate governmental broadcasting and nongovernmental clandestine broadcasting, etc. If nothing else, this volume makes evident the protean forms and sources of shortwave broadcasting, many of which will be unknown to the average listener.

Thousands of facts are presented here, making for an interesting indexing problem. In 19 pages of dense text, an appendix attempts to list all stations covered by country and, within country, by year. These pages must have required Job's patience to produce and proofread, but their usefulness immediately becomes obvious when particulars are wanted.

The *Listening* volume presents shortwave from the reception viewpoint, and is the one most beginning SWLers will turn to. There are detailed chapters on the shortwave audience, clubs, literature, listener programs, receivers, QSLing and computers.

The 100-page chapter on literature provides a good example of the inclusiveness of these volumes. In the subsection on "DX Newsletters" I count 19 titles covered, some with as few as 30 subscribers. Nevertheless, each entry includes a history of the newsletter, naming the major players, inclusive dates of publication, etc. Not content to let the matter rest there, the author follows up with "Other Newsletters," and these are further divided by geographical area and special topics–equipment, women, education.

Mr. Short Wave wrote to me in an email: ". . . no one is likely to walk this ground again any time soon." There are at least two reasons why this is so. First is the depth with which shortwave radio is covered in these three volumes. Second is the changing nature of shortwave; even if you don't read these books, do not deprive yourself of reading the conclusion Berg provides at the end of *Listening*–a bittersweet summary of the effects of computers and the internet on shortwave radio.

I have been an on again/off again shortwave listener for over half a century, but to this day I don't know the difference between the 80 meter band and the 40 meter band. I just turn the thing on and see what I hear, i.e., hear what I hear. And so, I believe, it is with many SWLers. As PARS- and AWA-member John W. Haught once said to me, "Shortwave radio is like a deep well. You never know what you're going to pull out of it."

These meticulous, first-rate volumes will be valued by shortwave enthusiasts and included in radio reference collections throughout the radio world. The set comes highly recommended.

Origins of the Vacuum Tube: An October 1964 Talk by G. F. J. Tyne
CD Edition. Tube Collectors Association, 30 pages, 2005.
AWA Journal, v46 n2, April 2005, p. 57.

Tyne's 1964 talk, introduced by Bruce Kelley, begins with the 1853 work of A.-E. Becquerel and ends with the Moorhead/de Forest tubes made just after the first World War. Tyne had original early tubes present at his talk, including some Edison-effect lamps and Fleming diodes. The audio of his speech makes evident his somewhat annoying verbal habit of starting a phrase or sentence several times before finishing it. Fortunately the habit did not carry over to his authoritative book, *Saga of the Vacuum Tube*, which is respected by collectors and historians.

The booklet contains black-and-white photographs time-keyed to Tyne's near-hour-long talk. Also included is a unique two-page table listing Western Electric developmental tube specifications.

Thanks to Jerry Vanicek for transcribing the original tape recording and producing this convenient, historical CD.

Pioneers of Electrical Communication
Rollo Appleyard, Macmillan, 347 pages, 1930.
OTB, v45 n3, July 2004, pp. 14-15.

These short biographical essays about Maxwell, Ampere, Volta, Wheatstone, Hertz, Oersted, Ohm, Heaviside, Chappe and Ronalds often include brief genealogies. Appleyard takes pains to show the frustrations and many false starts the pioneers experienced as they struggled to understand what could not be seen. He also frequently allows us to see the world as it was seen by the biographee.

Claude Chappe, who coined the word *telegraph*, was so stymied with his attempts to build a working electric telegraph that he completely abandoned it in favor of an optical (semaphore) system of signaling. That system ultimately encompassed 556 stations in France and was used throughout western Europe for the first half of the 19th century.

Ohm's law appears to us now as elegantly simple and obvious, yet to establish his work Georg Ohm struggled greatly against fuzzy definitions, unreliable instrumentation and simple human unwillingness to believe. He certainly could have made good use of a Fluke 177 with auto-range.

We easily assume today that Ampere must have basked in glory with his invention of electrodynamics, but Appleyard tells us he may have been more interested in psychology and chess than in his electrical work. (I sometimes like to fantasize about Herr Ohm (1789-1854), Monsieur Ampere (1775-1836) and Signor Volta (1745-1827) getting together on a rainy Saturday afternoon in 1825 in a little Swiss pub to "talk things over.")

And what of poor, humble Francis Ronalds? In 1816 he designed and constructed a working electric telegraph–frequently cited as the first–but now he is known primarily for the telegraphy library that he assembled, today housed by the IEE.

Appleyard provides many other insights and bits of information relevant to telegraph history. The only price the reader must pay is tolerance of a distinctly Victorian style, syntax and pace.

Practical Electronics for Inventors
Paul Scherz, McGraw Hill, 952 pages, 2007, second edition.
AWA Journal, v48 n3, July 2007, p. 37.

This is a neat, neat handbook. And one of the neat things about it is you don't have to be an inventor to use it. It's also useful to hobbyists, engineers, tinkerers, and anyone else who is terminally curious about electronics.

It's hard to believe that just one man wrote this book. Its big pages are information-dense, with intelligently-written text vying for space with graphs, tables, formulas, sidebars, waveforms and, generally, any information Scherz thinks you might need to accomplish your next project. Graphics are clear and appropriate and the schematics are beautifully drawn (by computer).

But first, are you a little rusty on theory? No problem. Scherz suspected so and provides a 275-page chapter appropriately titled "Theory". This section, revised and expanded from the first edition, amounts to a mini-course in electronics, and can be used as an introduction by the science-savvy beginner or as a refresher for the professional.

Scherz's attempt to make the information in the book useful to electrical engineers as well as hobbyists and tinkerers is largely successful, but the approach immediately brings up the subject of mathematics. There is plenty of it, but as Scherz advises: ". . . when the math in a particular section of this [theory] chapter starts looking ugly, skim through the section until you locate the useful, nonugly formulas, rules and so on, that do not have weird mathematical expressions in them." This constitutes a refreshing approach to an old problem.

There is essentially nothing here about antique radio, of course, but anyone working in the nitty-gritty world of contemporary electronics would come to treasure this reference book. A few of hundreds of practical examples: do you know how to operate a 1.5 volt light-emitting diode from the 120 VAC line? Need to know what a varistor does or what its symbol is? Or what if, like me, you use your oscilloscope only about once a year? In that case, you'll appreciate

the five-page section "What All the Little Knobs and Switches Do."
This handbook to today's electronics is readily recommended.

Pulse of Radio: The Autobiography of Sir Watson-Watt
Robert Watson-Watt, Dial Press, 438 pages, 1959.
OTB, v44 n3, August 2003, p. 60.

This is the shorter version of Watson-Watt's autobiography, contoured for American consumption. It condenses the 480-page British version, *Three Steps to Victory*.

Did Watson-Watt invent radar? Yes. He tells us so on page 319. (But see also his contradicting statement on page 11 of the *Journal of the Institute of Electrical Engineers*, volume 93, number IIIA, 1946). He also takes credit for the "Huff Duff" or *h*igh *f*requency radio *d*irection *f*inder and names himself as one of the inventors of operational research.

Robert Watson Watt (or Watson-Watt–the hyphen was added when he was knighted in 1942) had to be the Gladstone or Dickens of the radar world, or to put it another way, he was . . . well . . . full of himself. His writing style sometimes is a cross between 17th century Baroque and 19th century Victorian exuberance. Consider the following sentence: ". . . I have tried to avoid such distortions as are likely to arise when Watson-Watt argues with Watson-Watt about Watson-Watt, and when Watson-Watt reports the debate." Aside from the obvious egoism here, what does this sentence mean? Or this one: "But by the end of 1937 microwave radar seemed like a highly resistive, obstinately non-responsive cavity whose orifice carried an over publicized negative slogan by Dante." Watson-Watt enjoyed speaking and writing in broad metaphors, this last one referring to "Abandon hope all ye who enter here."

If you read only one of these radar histories–Buderi's *Invention That Changed the World* (see above) or Watson-Watt–which shall it be? If you want objectivity, the inclusion of the British and American sides of radar and what followed it, with lots of information about the major players, read Buderi. But if you like your radar history objective, British-oriented and with a large dash of *joie de vivre* thrown in, then you'll want to go with Watson-Watt (or Watson Watt). What?

Quintessence of Irving Langmuir
Albert Rosenfeld, Pergamon Press, 229 pages, 1962.
OTB, v44 n3, August 2003, p. 61.

Langmuir has always been on the periphery for me–glimpsed only occasionally and briefly somewhere out there on the edges of radio history. Then I discovered Rosenfeld's biography, buried in volume 12 of Langmuir's collected works.

This is the straightforward, unexciting telling of a life that, viewed one way, also appeared unexciting. Langmuir discovered science as a child and made it his life. Along the way he got married (once only!) and got a job with General Electric that lasted over 40 years.

But also along the way he traveled the world, conferred with some of the "biggies" in world government and science, proposed a theory of atomic structure, invented atomic hydrogen welding, thoriated tungsten tube filaments, invented the mercury vapor vacuum pump, became an air pilot, discovered how to induce rain and snow artificially, was a founder of the science of surface chemistry, published over 200 scientific papers, and indulged a myriad of hobbies, including mountain climbing, skiing, skating, and walking (he once walked 52 miles in a day). Oh, he also got a Nobel Prize in 1932 and, apparently, invented the grid leak circuit in 1913 (not to mention his 68 other U.S. patents).

As his biographer puts it: "Once he had gone to work for the General Electric Research Laboratory, which gave him complete freedom, and once he had married Marion Mersereau . . . he had the perfect job and the perfect mate. Against this backdrop of financial and emotional security, he simply spent the rest of his vigorous life on a smooth plateau of uninhibited scientific self-indulgence."

Or as Langmuir himself put it, "I have never spent a really unhappy day in my life." He was obviously never bored, either.

Radio Man: The Remarkable Rise and Fall of C. O. Stanley
Mark Frankland, Institution of Electrical Engineers, 356 pages,
   2002.
OTB, v43 n4, November 2002, pp. 45-46.

Here is a thorough biography of Charles Orr Stanley; it is also part
of the story of the radio and electronics business and industry in 20th
century Britain. As a young, self-made advertising executive,
Stanley, who knew next to nothing about radio, decided in 1928 to
buy Pye (pronounced *pie*), a small radio and instrument
manufacturer. The way Stanley closed the deal says much about him
and the future of his career. On the day of the sale, he carefully
waited until after the banks closed before he handed over his check.
This gave him the rest of the day to search for financing so the check
wouldn't bounce.

Stanley's daring and freewheeling (Americans might say *off the
wall*) business and management style frustrated some of his
employees but it fostered considerable creativity in others. By 1965
Pye was a conglomerate of 109 companies, including 39 based
outside Britain. Among their many products were radar equipment
and the proximity fuse, developed early in WW II. Britain struggled
to produce a few hundred such fuses per week. Against Stanley's
wishes, the technology for this device was "exported" to the U.S.; by
the end of the war the U.S. had manufactured 150 million fuses. Pye
also produced the famous No. 19 military radio, used worldwide, and
made Britain's first portable transistor radio in 1956. Stanley also
became one of the main players in breaking the BBC's broadcasting
monopoly.

It's interesting to try to find American counterparts to Stanley's life
and companies. His brashness in buying Pye reminds one of
Zworykin's cheek in building his house. When the owner of the land
on which Zworykin wished to build refused to sell, Zworykin simply
went ahead with construction. Eventually the owner agreed to sell.
There are obvious parallels between Stanley's bossiness (he could
tolerate no rivals in his companies) and Sarnoff and RCA. Pye's
chronically being strapped for cash, however, is paralleled by almost

any U.S. radio firm *except* Sarnoff's and Powel Crosley's. And it's easy to find U.S. parallels to Pye's weak, sloppy or devious accounting practices. Just watch today's business news. But neither the intent nor the result of "cooking the books" by current U.S. companies was ever duplicated by Pye. Stanley was a salesman first of all; accounting took a back seat. In the extreme, some Pye companies simply had no books to cook.

Some of Stanley's characteristics which made Pye successful also helped cause the spectacular crash of his career in 1966. This episode was so precipitous and so traumatized Stanley and his son John that Frankland begins his book with a chapter describing it. The chapter is titled "The Execution."

Every IEE History of Technology Series book reported on here so far has been beautifully and intelligently written; this one is no exception.

Radio Patent Chronology
David W. Kraeuter, www.lulu.com., Pittsburgh Antique Radio
   Society Publication 15, several pages, published about 2009.
AWA Journal, v52 n2, April 2011.

This book should never have been published. Or, to put the matter bluntly, what was the author using for brains?

Where to begin? How soon can we end? Just how often does this author think he can blithely bring something in from hard disc, rearrange it and then publish it again, claiming a new book has entered the world? He coyly states in the preface: "I'd like to provide this information as soon as possible to those clamoring readers who might crave it." No, no and no. Right from page one it becomes blatantly–shall we say *patently*?–obvious that the true motivation for publishing this book, like so many others being published today, is that of get-rich-quick, take-the-money-and-run, pure and simple greed. The *faux* noble goal stated in the preface is actually a cover to distract the reader's attention from the many over-wrought, emotional–dare we write *flights*?–of fancy the book contains, each masquerading as a statement of truth. But the only truth needed here is *Burn before reading,* conspicuously missing on page one. Spiders of the world rejoice–the book's about as effective as an Osage orange. Even silverfish will turn up their noses at this one.

You've all heard of books that couldn't be put down once the reader started reading; this one won't be picked up once it's put down. The book is sophomoric, soporific, puerile, incoherent, grammatically challenged and grandly dull. Even if this defecation of the mind were on fire inside a dynamite factory we still would be reluctant to micturate on it. The phrase "not worth the paper it's written on" was likely invented for this book. Reading it made us wish we were dyslexic. Surely even the printing press protested, the paper refusing to embrace the ink. Where was Goebbels the Book Burner when he was most needed?

Just think for a moment of the unfortunate editor who sacrificed his eyesight in proofreading this ostentatious treatise manqué. Did the author, even once, observe a moment of silence for the trees who gave their lives in vain for this shabby pretense of a production, or the electrons senselessly wasted in e-transmission? No wonder the Library of Congress put a minus sign in front of its ISBN !

Bloated as a week-old corpse, this whale of wordage surrounds a minnow of meaning; it should have been edited with a flensing knife. The $11.99 one is shamelessly asked to pay for this feeble effort would be much better spent as a donation to the Flat Earth Society, the Society for the Preservation of the Apostrophe, or, perhaps, the Andy Griffith Show Rerun Watchers Club.

We ask: the printing press was invented for *this*? Gutenberg must surely be spinning in his grave, even as we word-process these lines. This book is unequivocally *not recommended for anyone.*

Bottom line: landfill.

Note 1: the author was assisted in writing this review by William P. Keen, Edward E. Sweet, Jr. and John Alfred Taylor, each of whom wrote with "a deep sense of moral outrage". Also note that this review was published as close as could be to April 1.

Note 2: In his own defense, the author writes that "despite the negative tone of the review, I see that sales of the book have already soared into the high single-digit range."

Radio! Radio!
Jonathan Hill, Sunrise Press, 1986.
Pittsburgh Oscillator, v2 n1, March 1987, p. 4; reprinted in *OTB*,
v44 n1, February 2003, pp. 34-35.

Could you abide having a radio history book on your shelf that made no mention of A. Atwater Kent, Frank Conrad or RCA Radiola? Jonathan Hill's *Radio! Radio!* is such a book, but you're not with it for long before you don't care that the American side of radio history is generally left out. Hill's book is a 244-page chronology of the story of radio in England from its Victorian beginnings through about 1969.

It's fun to try to find the British counterpoint of the likes of Conrad (Captain Leonard F. Plugge?) or RCA (the BBC?) And just as the Hamilton Music Store of Wilkinsburg, Pennsylvania "sponsored" some of the first U.S. broadcasts, so Selfridge's Department Store in Oxford Street backed the first commercially sponsored program broadcast to British listeners. The year was 1925.

What was the British equivalent of the 01A "valve"? Probably Thomson-Houston Company's 'R' valve. It could be used as a detector, or in high or low frequency circuits. Hill reports that the tube's filament glowed so brightly that "some frivolous people, on occasions when they weren't using their receivers for listening-in, instead merely . . . turned up the rheostats high enough to provide light to read by or even to light the room. For those listeners who were annoyed by the bright glow of the R valve's filament, a Mr. R. F. Gordon of Weymouth had a remedy. He marketed a device costing 8d. "Rather similar in appearance and construction to a well-known family planning product, it consisted of a sleeve of black rubberised material which was rolled snugly over the valve, and being opaque, effectively shielded the listeners' eyes from the glowing filament."

In addition to the many details reported as charmingly as this there are the photographs. Almost a thousand of these, all monochrome, all in clear focus, and almost all captioned in great detail, listing make, model, date, dimensions, type of cabinet, type of circuitry, original price, etc.

Included in the book are nine appendices. The first, "The Golden Age of Loudspeakers, 1922-1930" makes a nice complement to Floyd Paul's *Radio Horn Speaker Encyclopedia*.

This book was obviously written with great care by someone who knew his subject well. (Hill's first book, *The Cat's Whisker, 50 Years of Wireless Design*, was published in 1978). The author also designed the art work on the dust jacket, and Sue Worral's rendering of it is stunning.

The only fault found with the book is the chronological organization of the index, which would have been more convenient for most users if it had been in one alphabet.

Radio Receiver Repair
John T. Frye, Howard W. Sams, 222 pages, 1959, second edition.
Pittsburgh Oscillator, v17 n1, March 2002, p. 14. Reprinted in *OTB*,
v43 n4, November 2002, p. 46.

We are all fortunate. Life is full of wonderful things–a snug bed in wintertime, the first thunderstorm of spring, the smile you caused on a friend's face, the realization that you're too old for military service. To this list radio servicepeople add the pleasure derived from the definitive repair of an old radio–the initial mystery and attendant questions, the hypotheses tested, the just-right degree of frustration and the final, demonstrably-rational fix.

We can increase our chances of arriving at this happy state by educating ourselves with radio how-to books, such as this one. Frye's repair book is highly recommended to those who have basic electronics knowledge and want to try repairing their own antique radios. Written in the mid '50s, it emphasizes transformer sets, AC/DC table sets and ac/dc/battery portables. Separate chapters cover auto radios, all-wave sets, and frequency modulation. A chapter was also added to this edition for those new transistor radios and printed circuits.

Frye has organized his book very conveniently by symptoms displayed by malfunctioning radios. Broadly, his book does for symptoms what the "Case Histories" section of Alfred Ghirardi's *Radio Troubleshooter's Handbook* does for individual radio models. Chapters in Frye's book include "The Dead Set, Tubes Light But No Sound, Only Slight Hum Is Heard, Only Noise Can Be Heard, Excessive Current Indication, Sets With Excessive Hum, Set Does Not Separate Stations, Sets That Whistle, Motor-boat, Etc., Noisy Sets, Sets With Distortion and Weak Sets." There are two chapters on intermittents.

The next time you get stuck while trying to fix an old radio or just don't know where to begin, flip this book open, match the radio's symptom to the appropriate chapter title, and start reading.

Radio Reception in the Twenties
Raymond M. Bell, UMI, 110 leaves, 1979.
Pittsburgh Oscillator, v1 n2, August 1986, p. 4.

Raymond Bell is the author of three informative books on the history of radio–*Radio Reception in the Twenties,* (written in 1979), *Television in the Thirties* (1973), and *Television Reception, 1948-1958,* (1980).

These books provide a detailed and fascinating account of early radio and television reception in and around Pennsylvania, written by a man who can name the exact day when he first heard a radio playing.

A companion pamphlet, *Sixty Years of Broadcast Reception*, is also available. It provides a thumbnail sketch of early radio, including important dates, notes on the advent of FM, details on how call letters were determined, and when and why various broadcast bands were assigned. For the curious, there is also an explanation why the modern television receiver has no channel one.

Radio Reception in the Twenties, University Microfilms International, (order number LD00261), 1979.

Television in the Thirties, University Microfilms International, (order number LD00019), 1973.

Television Reception, 1948-1958, University Microfilms International, (order number LD00386), 1980.

Radio Troubleshooter's Handbook
Alfred Ghirardi, Murray Hill Books, 744 pages, 1939–
Pittsburgh Oscillator, v7 n3, September 1992, p. 14; reprinted in
OTB, v44 n1, February 2003, p. 35.

We have the Douay, the Good News, the Geneva, the King James, the American Standard, and now the Ghirardi. I found a Ghirardi "bible" (so-called by some servicemen) at my last auction. I knew nothing about it and bought it more or less for the heck of it for $12. Then I got it home and started to browse. About an hour later I felt as if I held in my hands a book which listed everything anyone would ever need to know about any radio or service problem.

The Table of Contents alone takes a good while to read and digest. It lists 75 sections plus an index for this compendium of information aimed at the busy 1940s radio service person. A few section headings will suffice to give an idea of the scope of the book:

4. I-F Alignment Peaks of 20,816 Superheterodyne Receivers.
9. Gear Ratios and Scale Calibration Directions Required in Remote Tuning Controls for all Auto-Radios.
10. Routine Auto-Radio Installation and Interference Elimination Date Chart for All American Passenger Cars.
14. Recommended Replacement and Electrically-Comparable Batteries for 1250 Models of Portable Radio Receivers.
33. Pilot and Dial Lamps Characteristics and RMA Bead Color Code Chart.
55. Power Transformer High-Voltage-Winding Requirements for Operating Various Types of Rectifier Tubes.
75. Directory of Radio Manual, Handbook and Textbook Publishers.

But by far the most fascinating section of the Handbook is "Case Histories of Common Trouble Symptoms and Remedies for 4,820 Models of 202 Makes of Home- & Auto-Radio Receivers and Record Changers." Ghirardi says this section represents in condensed, tabulated form the accumulated servicing information gained by thousands of hours of actual experience in service work.

For each model this 405-page section lists problems common to that set, such as noisy reception, fading, broad-tuning, inoperative, etc. Possible solutions are described for each problem.

For example, when I was restoring a Philco 84 I noticed a loud hum in the audio even after the filter capacitors were replaced. After a lot of unsuccessful attempts to eliminate the hum I thought to look in the Ghirardi book. There for Model 84 opposite the heading "Loud Hum" were three suggested causes. The second one was "loose or corroded ground lug rivet for filament supply on 42 socket. Place an additional ground at this connection." By following these directions I eliminated the unwanted hum.

This is just the sort of guide needed by those of us who aren't too sure of what we are doing when we get inside a chassis and it has us stumped. It doesn't take many experiences like this with a service guide to make you want to keep it right there on your service bench forever.

Ghirardi bibles are rare and getting rarer. If you don't have one and see one at a sale, my advice is to say, "I'll take it." You can always ask the price later.

Radiola: the Golden Age of RCA, 1919-1929
Eric P. Wenaas, Sonoran Publishing, LLC., 475 pages, 2007.
AWA Journal, v48 n3, July 2007, p. 30.

Wowie, this is one fine book. Kudos to Wenaas and Sonoran Publishing for producing it. Do not wince at the price ($65.00)–it is worth every penny and more.

The book treats the Radiola story with knowledge, expertise and detail similar to that used by Ralph Williams in his treatment of Atwater Kents in the 1999 *AWA Review*. But unlike the Williams's book, *Radiola* is loaded with about 700 color photos of radios and equipment, advertising items and other close-ups. The photos remind me of the quality of photos in Ian Sanders and Carl Glover's *Tickling the Crystal* series on British crystal sets.

Nicely complementing the beautiful pictures is text containing enormous detail but no superficiality. Each chapter is extensively footnoted, with sources including, among others, the George H. Clark Radioana Collection now housed at the Smithsonian Institution. Wenaas states: "Every attempt has been made to locate and cite original documents of the era to ensure accuracy and minimize dependence on secondary sources." The photos, text, bibliography and footnotes amply substantiate this statement.

Wenaas does not present the Radiola *in vacuuo*, so to speak. There is plenty of background provided to show how Radiolas came to be. To give an idea of the thoroughness of the book: one would think that 400-some big pages would suffice to tell the Radiola story, but no, an additional 60 pages of appendixes (nine in all) are needed for such information as serial number data, release and end dates for receivers, model designation schema, advertising, etc. Detail, detail, detail–in the "Relative Availability of Radiola Apparatus" table Wenaas includes a column that gives a "Correlation of Metrics" figure for many Radiolas, and takes about a page to explain what this statistic means!

I cannot think of enough good things to say about this comprehensive work, and actually it deserves a much fuller review than I give here. How good is it? I spent hours with it in writing these few paragraphs and could find nothing to complain about, except one minor typo in the preface. (Also, the patent number at the top of page 105 should have been 1,664,192). *Radiola* will almost certainly increase the value of Radiolas everywhere, and that is a good thing, since it will provide further incentive for the proper restoration and preservation of these near-century-old artifacts.

No one in the United States is very far from a Radiola. Because of this book, I've already gotten out my RC, hooked it up, and will be doing some DXing tonight. You say you're not a Radiola enthusiast? Spend some time with this benchmark book and you will be.

Radios of Canada
Lloyd Swackhammer, Cober Printing Ltd., 164 pages, 2002.
OTB, v45 n2, May 2004, p. 25.

If the geese that often fly in ragged "V" formations above AWA's late summer playground in Henrietta, New York are Canada Geese–not Canadian geese–then by wacky analogy the radios in Swackhammer's book must be Canada radios. Whatever, this book has been needed if for no other reason than to make a start in plugging the gap in published information on this topic.

It is tempting to think of Swackhammer's book as a Canadian answer to Douglas's *Radio Manufacturers of the 1920's* for there are several obvious parallels–the 8 ½ by 11 format, alphabetical arrangement and use of contemporary reproductions, to name a few. But at 776 pages Douglas's work provides greater depth than this one.

The hook which keeps Swackhammer's book before us is the unfamiliarity of the contents. Ever hear of a Dominion Electro-home radio? How about a Dispro? A Pollock-Welker? Service any Walberts lately? Ever trade a Crosley Pup for a Spilsbury & Tindall Limited or a Penberthy Injector?

If you have serviced a Walbert, you may have used the schematic in the *Canadian Radio College of Canada* schematic books or the *Mallory Yaxley Encyclopedia*, either of which could be called the *Rider's* of Canada radios.

The two-part index to the book first lists product names, then companies and people. Many Canada radio enthusiasts will appreciate this book.

Restoring Baird's Image
Donald F. McLean, Institution of Electrical Engineers, 295 pages,
  2000.
OTB, v42 n2, May 2001, pp. 61-62.

In 1977 NASA launched two unmanned spacecraft, Voyager I and
II. They spent the '70s and '80s exploring the outer planets and are
now slowly (at about 35,000 mph) traveling out of the solar system.
Though NASA and the Voyagers chat with each other regularly, it
takes about half a day for the signals to travel in either direction. You
can check up on the Voyagers' progress by visiting:
         http://vrapter.jpl.nasa.gov/voyager/voyager.html.
  Peripatetic astronomer Carl Sagan convinced NASA late in the
game to include aboard each spacecraft a gold-coated copper
"phonograph" record. The records each contain 118 photographs of
our planet and its inhabitants, greetings in about 60 languages
(including one whale language) and about 90 minutes of "the
world's greatest music," this last including Chuck Berry singing
Johnny B. Goode.
  Science assures us that someday the sun is going to burn out or
explode. Either event will likely cause considerable discomfort to
folks living here on earth. But not to worry–Beethoven's matchless
Cavatina will continue hurtling blithely on through space to   . .
where? Sagan estimated that in the cold vacuum of deep space the
phonograph records should last for "about" a billion years. Some
"day" or some eon, somebody–or some thing–may discover one of
the Voyager discs and learn how to play it, using the stylus and
cartridge that NASA so thoughtfully provided.
  If you just can't wait to see how the Voyager story ultimately plays
out, you might want to consider the experience of Donald F.
McLean, whose recent work decoding John L. Baird's 1920s video
recordings has parallels in the long-playing Voyager saga. These
parallels are obvious, and in fact McLean includes a few pages about
the famous spacecraft near the end of his book. He concludes that
whoever may find the Voyager phonograph records "will then have
the task of restoring the age-old video, using techniques that may not

be all that different to those used for Baird's Phonovision."

The clever title of McLean's book permits two interpretations, literal and figurative. The literal work McLean did in "bringing back" and computer-enhancing television images from Baird's 1920s Phonovision records was outlined in McLean's article beginning on page 12 in the last (February 2001) *OTB*. If you liked the article, you're pretty much guaranteed to like his book, which elaborates in great detail (too much detail to be reviewed here) the technical side of recapturing the images.

But McLean also makes a strong case for restoring Baird's image and stature in television history. He rightly cautions us against thinking of Baird as having the "wrong" technology. To do so would be similar to thinking that Edison's cylinder phonograph was the wrong technology in an age when Emile Berliner's flat discs had taken over. Baird didn't have the wrong technology; he had the technology of the 1920s and '30s.

Also, though Baird may have started out using "sealing wax and string" equipment, he continually upgraded and improved his apparatus. (It is not an easy thing to convince the British patent office to issue a patent; Baird had 175 of them). Baird personally demonstrated a mechanical color television system a few years before he died. A key component was a 12-inch diameter 20-facet mirror drum that rotated at 6,000 rpm, subjecting the mirrors to about 2,000 times the force of gravity. No sealing wax or string here.

We are in Carl Sagan's debt for producing the records on board the Voyagers and we are in Donald McLean's debt for decoding and enhancing the Baird Phonovision discs, thereby also restoring Baird's image today.

For the story of the Voyager records, see *Murmurs of Earth: The Voyager Interstellar Record* by Carl Sagan and others, published in 1978 by Random House.

Rise of Radio, From Marconi Through the Golden Age
Alfred Balk, McFarland & Company, Inc., 350 pages, 2006.
AWA Journal, v47 n3, July 2006, p. 41.

Balk sets the Golden Age of radio at about 1926-1952. I'm a radio technology and hardware junkie, so I assumed that once I'd read the initial chapters about radio's invention I would lose interest in the rest of the book, which deals mainly with Golden Age performers, programming, economics, politics and social trends. But that did not happen, for at least two reasons–first, Balk knew or learned his subject very thoroughly and second, he is an accomplished, insightful writer (he has written or edited half-a-dozen earlier books).

The book can be read enjoyably straight through or one can sample from the various genres laid out in about 30 concise chapters– comedy, adventure, children's radio, educational programs, the soaps, plays, big bands, quiz shows, talking heads, etc.–all thoroughly researched and documented, but with some occasional gossip thrown in as leavening.

Balk's last chapter, "A Legacy Lost," makes for some pretty sad reading. It chronicles commercial and public radio's decline in the last half century and can be succinctly summarized with a few catch phrases–"money talks," "whatever the market will bear," and "lowest common denominator" come readily to mind. (When was the last time you deliberately listened to an AM broadcast?)

Despite the book's somewhat generic title, it is limited almost entirely to U.S. radio history.

Sears Silvertone Catalogs 1930-1942: Complete Price Guide to Antique Radios
Mark V. Stein, Radiomania Books, 239 pages, 2001.
OTB, v42 n3, August 2001, p. 29.

The book consists almost entirely of full-size reproductions of the radio pages of Sears catalogs from 1930 through 1941. Note that the title can be misinterpreted–detailed coverage does not include 1942. This is the latest in a series of price guides by the publisher, earlier guides including pre-war consoles and three volumes of tabletop radios.

Suggested current values for each set are included in an eight-page table at the back of the book, but this is not the most valuable information provided. One suspects the price guide (so prominent in the book's title) may have been included to enhance book sales. A manufacturer source code table rounds out the book. The table identifies, from the chassis number, which company made any specific set for Sears. About 35 companies are listed.

The real value of the book lies in the unstated sociological, historical and economic lessons provided. Here is a subtle picture of several aspects of American life in the years of the Great Depression, particularly life on the farm. In 1935, the year in which the Rural Electrification Administration was established, only about 10% of U.S. farms had electricity. For that reason a large number of the radios covered here are battery-powered, including a surprising number manufactured in 1941.

Batteries and their foibles, particularly periodic charging of "A" batteries, must have generated a lot of talk among the owners of these radios. No one wants to transport heavy batteries to be recharged, but this was done on a regular basis. Sears also sold several devices for home charging by the farm set owner. One of the more interesting of these was the Silvertone Air-Charger, a wind-driven propeller-generator combination designed to be mounted on a rooftop. It required an average wind velocity of 8 to 12 miles per hour to effectively keep a 6-volt storage battery charged. If you lacked adequate wind power, perhaps you needed the Gas-O-Power gasoline-operated self-starting battery charger. This was itself started by a 6-volt storage battery.

By the late 1930s Sears was offering radios powered by just one battery. These contained vibrator-operated power supplies, just like in the car radios that we hated to service in the 1950s. As early as 1933 Sears was offering a car radio, and that catalog also boasts the "newest thing in radio," a Silvertone portable. The smiling, slender, young woman pictured carrying one of these radios appears not to notice that it weighs "only" 32 pounds.

By 1937 Sears had discovered advertising testimonials. So we have pictures of Mr. Chester E. Nowell of Georgia, Mrs. Steve Polasek of Ohio and Mr. H. W. Adams of Delaware, each singing the praises of the Sears radio. We are assured that their letters are on file in Sears' offices. We know roughly where these people lived, but just what motivated them to write? By 1941 Mr. J. L. Hammond (pictured in bib overalls) of Sanford, Maine gives us about 250 words of praise for his Sears' set. Were these testimonials genuine, contrived or somewhere in between?

I have never seen a Sears radio that I found pleasing to the eye, and I wonder if this aspect of their design was ever addressed by the company. Perhaps the way the set looked was thought to be last in the mind of the consumer, or, as was recently suggested to me, perhaps Sears may have been deliberately trying not to offend the aesthetic sense of its customers. We may never know.

This will be a must-have for Sears' collectors and enthusiasts, but others will also be able to appreciate the rich detail and nostalgic takes presented here by the dozens.

70 Years of Radio Tubes and Valves
John W. Stokes, Vestal Press Ltd., 247 pages, 1982.
Pittsburgh Oscillator, v2 n3, September 1987, p. 6.

Here is another of those books for which the hobby is greatly indebted to the author. Stokes's book constitutes a first-class professional achievement in all respects. From the beautiful color frontispiece of a de Forest Audion to the glossary and index, the book seldom disappoints.

The author has been interested in radio since 1929. That was the year in which he heard his first crystal set. During his lifetime he has amassed a large library on the subject, and has used it as source material in his writing.

The book is extremely detailed. Witness Stokes's introductory remarks titled "Generic Vacuum Tube Terminology," in which he analyzes the Greek etymology behind the designations *diode, triode, tetrode, pentode, hexode, heptode, octode* and *nonode*.

For each new grid invented and developed, Stokes adds another chapter to his book, and covers the item thoroughly. There are also separate chapters for double-filament and multiple tubes, contacts, frequency changers, miniaturization, bulbs, electron ray tubes, power rectifiers, transmitting tubes, metal envelopes, and other aspects.

The chapter on contacts alone provides a good example of the range and type of information provided. It consists of 20 pages of highly detailed information about tube pins, bases, and sockets in Europe and the United States from the 1920s through the 1940s, and includes photographs, drawings and reproductions of contemporary advertisements.

Perhaps most interesting to the antique radio enthusiast is the chapter, "Tube Collecting as a Hobby." Here Stokes details such items as Japanese tubes of the 1930s, Russian and Italian tubes, "private brand" and fake tubes, and the ever-popular Japanese UX227, which, according to the author, doesn't exit. (It's difficult to fault an author of a book on tubes who spends time writing about a tube which doesn't exit). The chapter closes with tips on displaying and marking tubes.

Throughout the book hundreds of tubes are pictured and identified. In addition to British and American tube makers, French, Canadian, and Australian firms are also included.

Signor Marconi's Magic Box: The Most Remarkable Invention of the 19th Century and the Amateur Inventor Whose Genius Sparked a Revolution
Gavin Weightman, Da Capo Press, 312 pages, 2003.
OTB, v45 n1, February 2004, p. 45.

"Once upon a time in a land far, far away . . . ." So might Weightman's book have begun, for it is nothing less than a storybook–but for adults. There are no end notes or bibliography and very few footnotes; all these scholarly trappings would have gotten in the way of the stories Weightman wants to tell. It is nevertheless obvious from the text that Weightman read widely and deeply about his subjects before starting his narrative.

Weightman's cast of supporting characters is extensive and helps the reader see Marconi in cultural and historical perspective. So we have the story of Maskelyne brashly but correctly demonstrating that Marconi's wireless communication system was not as private as he liked to say it was. "Crazy" deaf Heaviside makes a brief appearance in the narrative, and two chapters are devoted to Jack Phillips and Harold Bride, and Harold Cottam, who were radio operators on the Titanic and Carpathia, respectively.

Weightman's telling of the story of Marconi's meeting with Bride and Cottam after the Titanic's disaster is nothing short of poignant. (Phillips did not survive the disaster). Marconi was not recognized by the authorities and had to talk his way past guards before he was allowed to board the Carpathia. Marconi was originally scheduled to be on the Titanic's maiden voyage, but had to cancel due to schedule conflict. Nevertheless, the public came to see Marconi as a hero of the disaster because the Titanic had been fitted with radio equipment made by the Marconi company.

The public today naively sees Marconi as the inventor of the modern radio, but Marconi himself saw radio mainly in terms of private, long-distance, coded point-to-point transmissions rather than voice or music broadcasting, even after the public broadcasting radio craze of the early '20s. And many others had a hand in the invention of radio; indeed Marconi "ran scared" for years in his early career, often worrying that other radio inventors–Fessenden, de Forest, etc.–would eclipse him, particularly in distance transmission.

Although Weightman obviously admires Marconi, he does not shy from presenting less positive aspects of Marconi's life, such as his overdriving ambition, his neglect of family life, and his support of fascism in the 1930s.

These beautifully-told stories are highly recommended for all lovers of technobiography.

Sir Oliver Lodge, Psychical Researcher and Scientist
W. P. Jolly, Fairleigh Dickinson University Press, 256 pages,
   1975.
OTB, v43 n3, August 2002, pp. 53-54.

Lodge's long active life (1851-1940) was divided among very many interests, one of which was the newly-founded study of electromagnetic waves. Just as waves in a fluid, such as water, could not exist without the water, so too it was thought there must be a medium by which electromagnetic waves existed. This mysterious substance was called ether, and Lodge studied it during much of his adult life.

He devised an experiment to determine whether a large rotating object would pull or drag the surrounding ether along with it. The experiment used two heavy, closely-spaced one-meter-diameter steel discs, each spinning at up to 4,000 rpm. Some danger was involved. "Since the optical measurements were made with [Lodge's] head . . . in the same horizontal plane as the spinning discs, the nature of the possible disaster was uncomfortable to contemplate." Though the experiments lasted for years, Lodge survived and no indication of ether drag was found. Einstein would later use the results of Lodge's work in the formation of his famous theories.

Think of Lodge the next time you tune in a station on your radio or TV. His most important contribution to radio circuitry was the resonating circuit that allows a radio to select just one of the many stations that may be available to it. Some people think that Lodge, rather than Marconi, should be credited with the invention of radio. Indeed the famous British 7,777 tuning patent was granted to Marconi in 1901, but Lodge got his less-than-famous tuning patent in 1898 (British 609,154).

Lodge won credit for his work only through a trial, which extended his patent through 1918. Marconi's company bought Lodge's tuning patent after the trial. Jolly portrays Lodge as a kind and gentle man, and says that Lodge bore no feeling of ill will toward Marconi. He could, however, bridle at the term "Marconi waves." If an eponym were needed, Lodge much preferred "Hertzian waves."

For other views of Lodge's contribution to radio, see the collection of essays edited by Peter Rowlands and J. Patrick Wilson titled *Oliver Lodge and the Invention of Radio* (Liverpool: PD Publications, 1994), reviewed here on page 121.

Of Lodge's many non-radio interests one of the more bizarre was spiritualism, which in Victorian times meant communication with the spirits of dead people, especially relatives. (Lodge certainly had ample opportunity here; his father had 24 brothers and sisters). Another long-time investigator of spiritualism was William Crookes, who was one of the inventors of the CRT, the discoverer of thallium, and the inventor of the radiometer.

Lodge believed he had confirmed or demonstrated communication with the dead, but realized he'd have an extremely difficult time convincing the scientific community. The general public was another matter. Lodge's book *Raymond*, about his attempts to communicate with his son who had been killed in the Great War, caused something of a sensation and became what would today be called a bestseller. The popularity of the book was doubtless caused in part by the grief of the many people who had lost relatives in the War.*

At his death Lodge left a deliberately unfinished experiment designed to test or demonstrate the truth of spiritualism. He left a packet of seven sealed envelopes of different sizes, each envelope containing the next smallest, and each bearing clues as to the message in the smallest envelope. That message was known only to Lodge. These clues were to help mediums, people who were supposedly sensitive to communication with the dead, in communicating with Lodge's spirit. The Second World War interfered somewhat with the experiment, but the envelopes were finally opened successively between 1947 and 1954. Jolly reports that the results of the experiment were inconclusive "to the uncommitted observer."

Jolly repeatedly states that Lodge brought to his spiritualism studies the same rigorous scientific methods he used in his studies of electromagnetic waves. But the question remains: how could men of Crookes's and Lodge's obvious intellect and devotion to reason be so caught up in spiritualism? I don't know the answer, but for a thought exercise of our own I'll end with David Hume's famous sentence, "Reason is, and ought only to be, the slave of the passions."

* *Raymond* is reviewed here on page 212.

Snapshots in Time: An Electronics Anthology
Ted Depto, P.E., Vintage Radio and Electronics, 87 pages, 2008.
AWA Journal, v50 n3, July 2009, pp. 67-8; reprinted in *Pittsburgh Oscillator*, v24 n3, September 2009, p. 13.

Having edited the *Pittsburgh Oscillator: Journal of the Pittsburgh Antique Radio Society* for the past dozen years, Depto decided it was time to put some of his many ideas about radio into book form. Editing was not Depto's only qualification for writing this book. He served in the Signal Corp during the Korean conflict, has a degree in electrical engineering technology from Penn State University, and operated an electronics business in the '50s, '60s and '70s. His book covers so many varied aspects of electronics repair, collecting and restoration that it is difficult to classify and even more difficult to review.

Both in his writing and in his service work, Depto investigates where others don't, and often sees things others overlook. Two examples will elucidate. When Depto came across a Delano Sheraton TRF radio from the mid '20s, he couldn't rest until he tracked down the manufacturer, the Modernola Company in Johnstown, Pa. His research uncovered that the company was incorporated in 1874 and went out of business in the late 1920s. Though the company has been defunct for 80 years, Depto still includes a photograph of the company's factory as it appears today.

As for having an eye for detail, here is what Depto wrote in his "Tid-Bits" column in the June 1999 issue of the *Oscillator:*

> *If you have ever worked on an Atwater Kent Model 20, either the compact or the big box, you may have discovered . . . that underneath the Atwater Kent nameplate fastened to the front panel with bent over fasteners, there is an interesting phenomenon. Behind this small plate is a piece of folded paper which says, "Not Guaranteed".*

This "phenomenon," as the author calls it, seems to me to border on the radio kinky. Does anyone know the reason for this presumed AK practice?

Other unusual features covered in the book are Airline's 1936 "Movie Dial" radio, which projected the call letters of the received station onto a screen on the front panel, radios built into models of ships, and a 1936 Globe "Navigator" radio, housed inside a globe representing the earth–to tune the radio, just spin the ten-inch globe. These radios may all be manufacturers' responses to a depression-era market that was also saturated. By the mid '30s, most homes already included one requisite "family" radio.

Depto himself gives a hint about the nature and style of these essays in his foreword: ". . . not too technical, not too earth shaking, just right for an evening of reading". I would add that they are happily rough-hewn and eclectic, and always include some unexpected nugget of interest. We who enjoy this "down-home" approach to radio history are happy to learn that a second volume of Depto's radio essays is nearing completion.

SOS Korea 1950 Illustrated
Raymond B. Maurstad, Beaver's Pond Press, 396 pages, 2003.
OTB, v45 n1, February 2004, p. 46.

   This is largely the story of American involvement in Korea from the end of WW II through the beginning of the war in Korea in 1950. The story is told through personal biographical accounts of about 100 people. The author/editor was a professional radio operator, and the book is included here because of several chapters covering the role of amateur and military radio communications in Korea at the time. Particulars are supplied for the truck-mounted SCR 399 high frequency radio for telegraphy. Includes about 140 monochrome and color photos of contemporary Korea buildings and scenes.

Stay Tuned: A History of American Broadcasting
Christopher H. Sterling and John M. Kittross, Lawrence Erlbaum
  Associates, 975 pages, 2001, third edition.
OTB, v43 n2, May 2002, pp. 52-53.

Here is American broadcasting history in a 4.5-pound "nutshell." The first (1978) edition of this book was 562 pages long and the 1990 edition was 705. This is obviously an ongoing project with much thought and work between editions. Thoroughness is the touchstone here, and the present edition is in fact several books in one.

Stay Tuned is designed to be used as a text for a college or university course, but it is so thorough in its coverage and so well-organized that it can be used several other ways. The text can be used as a chronology (what happened and when) or as a history (what happened and why). In fact the book contains a separate 20-page chronology. But because of the book's extensive indexing and thoughtful organization, it can also be used as a single-volume encyclopedia of its subject. In addition to the usual table of contents, an "alternate contents" section lists information by topic, and each chapter is preceded by an outline and followed by a list of selected further readings. Used this way *Stay Tuned* can be seen as a companion to (or competitor with) Godfrey and Leigh's *Historical Dictionary of American Radio* (Greenwood, 1988) or Sies's *Encyclopedia of American Radio, 1920-1960* (McFarland, 2000) or Lackmann's *Encyclopedia of American Radio* (Facts on File, 2000).

A whopping 182 pages are devoted to four appendixes. The chronology is followed by a glossary with word "definitions" sometimes over a page long, making it a mini-book in itself. This is followed by a section of historical statistics (fun to dip into at random), and a 54-page selected bibliography. So another use of the book is as a guide to the literature. (But see also Sterling's *History of Telecommunications Technology: An Annotated Bibliography*, reviewed in the November 2000 *OTB*). The bibliography also contains three mini-appendixes: a list of web sites, museums, and libraries and archives. Inclusion of so much "value-added" material

159

may tempt libraries to keep this book in the reference section, rather than circulating it.

Stay Tuned is not for the casual hobbyist and it's likely hardware enthusiasts will be disappointed despite the color photos on the covers, but if you're looking for detailed, balanced introductions to virtually all other aspects of American broadcasting this will be a welcome find.

Steinmetz: Engineer and Socialist
Ronald R. Kline, John Hopkins University Press, 401 pages, 1992.
AWA Journal, v48 n1, January 2007, pp. 17-19.

Kline's biography of Steinmetz is appropriate for inclusion here, not only because it was published in the same year and by the same press as Alexanderson's biography. Steinmetz, the Wizard of Schenectady, was Alexanderson's mentor at General Electric. Though Steinmetz had nothing to do directly with the invention of radio (his main work was in AC power engineering), he certainly had a lot to do with the making of most electrical engineers in the twentieth century and today. This was largely accomplished through his leadership in the American Institute of Electrical Engineers and through his publication of ten-or-so books which have been called "classics of electrical engineering" and have been translated into many languages.

Can you see a link between electricity and socialism? Russian revolutionary Lenin thought he could when he wrote his famous dictum: "Communism is socialist power plus the electrification of the whole country . . . ." Steinmetz wrote to Lenin offering his services. He also gave numerous speeches and produced many publications expounding his ideas about the "engineering of society". The second half of Kline's book is devoted to the story of Steinmetz's socialism and to explicating the many myths about Steinmetz, who was, by the end of his life, more appropriately referred to as the Oracle of Schenectady.

"Study of the Operating Characteristics of the Ratio Detector and Its Place in Radio History"
Edwin H. Armstrong, *Proceedings of the Radio Club of America*,
  v25 n3, November 1948, pp. 1-20.
OTB, v43 n2, May 2002, pp. 53, 63. Reprinted in *OTB*, v43 n3,
August 2002, pp. 54-55.

Just as there are seminal books in the electrical engineering field, so too there are seminal papers. This paper by Armstrong, though not of prime value, is still important for two reasons. First, it is his public refutation of the originality of the ratio detector circuit, and second, it gives us an all-too-rare glimpse into his humanity.

Armstrong begins his paper with an historical precedent. His 1912 invention of the regenerative circuit had become well-known by late 1913. Early in 1914 Lee de Forest announced his invention of the ultra-audion circuit, which he claimed was not a regenerative circuit but achieved the same results. But by simply re-drawing de Forest's ultra-audion circuit to include coupling and the inter-element capacitance of the tube, Armstrong was able to show that the ultra-audion circuit was in fact his own regen circuit in disguise. Regardless, the U.S. Supreme Court (if not the electrical engineering community) ultimately found in favor of de Forest in 1934.

Then Armstrong turns to the main subject of his paper. In the early 1930s he had developed and patented his system of noise-free wide-band FM radio using limiter and discriminator circuits. Acceptance of Armstrong's FM inventions occurred not nearly as fast as with the regen circuit, which could quickly be incorporated into receivers and required no modification of the transmitting circuits. FM, however, required a complete new system. For years it must have seemed to Armstrong that no one cared or was listening when he promoted his new noiseless system.

But obviously someone was listening, for by 1947 David Sarnoff and RCA engineers, particularly Stuart Seeley, introduced a circuit which purportedly did the same thing as Armstrong's FM circuits, but, again, in a unique way. Seeley's circuit was called the ratio detector, and RCA was able to get the circuit patented despite

Armstrong's earlier patents. (See my "Seeley Bibliography" in the *Pittsburgh Oscillator*, December 2001).

For Armstrong it was to be, as the redundant saying goes, "*déjà vu* all over again." Once more he found himself on the defensive. And once more, by a logical step-by-step redrawing of the ratio detector circuit, he was able to show that it was in fact merely a variation of his own original limiter/discriminator design. But when this paper was published in 1948 Armstrong still faced years of frustrating, expensive litigation with RCA over his FM circuits.

Note by the title of the paper that Armstrong wished to show the place of the ratio detector in radio history. He does that none too subtly on the last page. Just who might Armstrong have had in mind when he ended his objective, technical paper with this surprising (but still carefully footnoted) fillip?

> *That prophecy* [made by Armstrong] *was that the day would surely arrive when the direction of engineering by the members of the legal profession would come to an end, because the unholy mess that they had made of radio would soon be apparent to everyone. The writer* [Armstrong] *predicted that engineering would again be directed by engineers, and he even ventured to think that the day might arrive when some highly successful executives would come to believe that there was something after all to the text of the Eighth and Ninth Commandments,[3] stating that in case the audience could not immediately place them by number that they were "Thou shalt not bear false witness against thy neighbor" and "Thou shalt not steal."*

> [3] *(Protestants numbering, generally).*

TCA Data Cache
DVD-ROM plus eight-page descriptive booklet. Tube Collectors
   Association. Access via Adobe Reader version 8.1, 2009.
AWA Journal, v50 n4, October 2009, pp. 27-28.

The dictionary says *cache* (sounds like money) means a hiding or
storage place. This latest production by the tube people (a.k.a. Tube
Collectors Association) certainly fits the second part of the
definition. On this single DVD can be found way over 10,000 pages
of highly-specific information about active electron devices. (It is
nigh impossible to determine an exact page count since the
information originated from many different sources and the pages
have not been numbered in one sequence).

A few caveats at the beginning: do not think of this DVD as "a big
tube manual." Though tubes are the main feature, it also contains
information on many transistors and other solid state devices. And
note that the information contained on the disc was not originally
meant to be released to the public, at least not in the forms presented
here. That fact can be a negative or a positive, depending on the
needs and motivations of the disc user. See the "Capabilities" and
"Limitations" sections provided in the eight-page booklet.

The information comes in three groups–that produced by tube-
manufacturing-giant RCA on about 2,800 tube types, about 7,000
records from 1947 onward from JEDEC, the Joint Electron Devices
Engineering Council, and a collection of 43 photographs depicting
production of a radar oscillator tube during WW II. For a fuller
description of the specific types of material on the DVD, visit
www.tubecollectors.org and click on "Notices", or see pages one and
two of the April 2009 issue of *Tube Collector*.

Of course, for pinout info you'll just grab your *General Electric
Essential Characteristics* or *Tung-Sol Base Connections*. But for
much greater detail, including lots of business and manufacturing
esoterica, you'll want to consult this disc. I went right for the type 80
tube, whose operation I think I know best–AC in, DC out. Even
though the RCA cache generally covers information produced years
after the 80 was devised, I still found three pages of 80 information

here. I next visited the JEDEC cache, where I found nine more pages of information–way more than I'll ever need to know, but still somehow real nice to have.

The 43 Eitel-McCullough photos are icing on the cake. Anyone who has never seen a tube manufacturing process will find these step-by-step views interesting, made even more so by their being 65 years old. Don't look for automated, assembly-line slickness here. Most operations are carried out by human hands, performing the same procedures over and over. OSHA mandates are glaringly obvious by their absence, but . . . we were at war. No . . . we were at War.

Having so much information on this disc required close attention to navigation and indexing tools. I found that some pages of JEDEC information about the 80 had been accidentally copied upside down. The "rotate view" command in Adobe Reader quickly corrected the problem. Its magnifier tool was also useful for fine detail.

Finding info on the 80 in the RCA cache was no problem, since the records there are arranged by tube type number. I quickly found the 80 by clicking on the folder labeled *35DZ8* [to] *117Z6GT* and then "paging down". Finding stuff on the 80 in the JEDEC cache was another matter. Information there is arranged by JEDEC release number, which will almost always be unknown by the cache user. James Cross comes to the rescue with his many eponymous indexes, one of which is from type number to JEDEC release (or folder) number. A table listing operations performed on the many JEDEC folders prior to digitizing them, such as eliminating blank and duplicate pages, is a nice indicator of the thought and preplanning that went into this major project. Congratulations to the tube collector guys in general for this gargantuan compilation of "behind the scenes" information.

I have always liked books. I think that they neatly "trap" information and can make it easily available within minutes to anyone who wants it. So too with this half-ounce DVD. Think of it–3.45 gigabytes of information! Microfilm–the cat's meow in the heyday of my career as a librarian–must surely be embarrassed by this new format.

So let's see–$50 divided by way over 10,000 pages equals way less than half a penny per page of information. When I consider the man-hours that went into the production of those tube and solid state devices, and the further man-hours that went into the writing of those 10,000+ pages, $50 sure sounds like a bargain to me.

Television and Me
John L. Baird, Mercat Press, Ltd., 148 pages, 2004.
OTB, v45 n4, October 2004, pp. 59-60.

One might think that Russell Burns's 417-page *John Logie Baird: Television Pioneer* and Antony Kamm and Malcolm Baird's 465-page *John Logie Baird: A Life* would say about all that needs to be said about Baird. But this publication of Baird's updated autobiography is a happy event for two reasons: first, it's always best to get your information from a primary source, if possible, and second, Baird provides some insights into his life and work that his biographers did not include. For example, on page 54 Baird writes:

> *The ideas regarding television in 1925 are surprising today when television is taken as a matter of fact, and the method by which it is accomplished is also now regarded as the obvious method. In 1925 it was not so. In fact it was the very reverse.*

These are telling words from the man who invented television using spinning cardboard discs and wound up producing the world's first cathode ray color picture tube.

Baird's frank descriptions of his attempts to enter the business world prior to his work on television are, in places, hilarious. For examples, see his descriptions of making guava jam in the Caribbean or his experiences with the Baird Under-sock and "pneumatic" walking shoes. What would he think of today's walking shoes with compressed air in the soles?

Baird wrote with candor and good humor throughout his autobiography, so it is doubly disappointing that he omitted his thoughts and feelings at the time he abandoned mechanical television for a cathode ray system.

He was an electrical innovator of the first rank. This fact does not prevent his life from being read as a study in frustration and perseverance. While continually inventing he saw his way through many personal and professional setbacks, until finally he was poised for success with his various companies and held a monopoly on color television. Unfortunately he arrived at this position in 1939.

Tempter
Norbert Wiener, Random House, 240 pages, 1959.
OTB, v42 n4, November 2001, pp. 30-31.

Fiction is rarely reviewed in this column, but for the intriguing combination of Norbert Wiener (1894-1964), Michael Pupin (1858-1935), Oliver Heaviside (1850-1925) and George Campbell (1870-1954) we must make an exception.

Norbert Wiener, a mathematician and one of the founders of cybernetics, was a genius who "spoke many languages but was not easily understood in any of them."[1] Obtuseness is not the case, however, with Wiener's posthumous book, *Invention: The Care and Feeding of Ideas*, in which he is almost brutally cogent and logical.

Many *OTB* readers would be interested in reading *Invention*, though it is not appropriate for review here. But in about ten pages of the book Wiener outlines the historical, legal and moral relationships that developed among Oliver Heaviside, Michael Pupin and George Campbell over the invention of the telephone line loading coil.

By about 1900 the telephone industry wanted to expand from intra-city service to inter-city service, but there was a technical problem. Long before telephone signals became unusable because of attenuation in long distance lines they became unusable because of distortion. Heaviside, Pupin and Campbell all worked on this problem. The general solution, contrary to standard opinion, lay in "loading" the lines by adding inductance coils.

The solution having been found, the inevitable questions arose: who was first and who gets the money? (This is not always the same person). Almost certainly priority must go to Heaviside, but he did not patent his work. Pupin did patent his work, but it may have been preceded by Campbell's. Campbell's employer, AT&T, nevertheless bought Pupin's patents for about a half-million dollars. The ploy was that buying patents from an "outside" researcher would look better in a patent dispute than if AT&T used the work of one of their own employees.

Wiener became so interested in the Heaviside/Pupin/Campbell story that he wrote a novel about it. Wiener was certainly a better mathematician than a novelist. In *Invention* Wiener described Heaviside as "an undersized, hungry, deaf, cantankerous little electrician." But in *The Tempter*, Heaviside, as the character Cedric Woodbury, is now described as "a cranky old codger with a tongue like a black-snake whip."

One of the reasons historical fiction is so interesting is that it must be based on fact. Unlike pure fiction, in which anything can happen (as long as it *could* happen), historical fiction requires that not only could it happen but that it did happen–at least in part. But just how factual must historical fiction be? Therein lies the rub–and the interest. We like to think, for example, that all sentences appearing in, say, the *OTB* are true, but in historical fiction the verity of any given sentence can be suspect. This situation creates tension, and tension creates interest.

Enough literary theory–what about Wiener's novel?

In the novel Pupin becomes Diego Dominguez, Campbell becomes Watman, AT&T becomes the Williams and Albright Company and the loading coil becomes a feedback control device. The novel, of course, allows Wiener to say and imply many things about Heaviside, Pupin and Campbell that he could not do in real life. But Wiener comes down pretty hard on Pupin both in real life and in the novel. In *Invention* he describes Pupin's moral position as untenable, and says that as a result Pupin's autobiography can be interpreted as a "cry from Hell." In *The Tempter,* Wiener says of Pupin (Diego Dominguez), "At the cost of his soul, he had accepted the wealth and power secured to him by the modern magic of science . . . . In fifty years or even in twenty-five he would become one of those false heroes of whose exploits history is full . . . ."

So how much of Wiener's novel are we willing to believe is historical fact? And just who did invent the long distance telephone line loading coil–Pupin? Heaviside? Campbell? You can read any or all of the publications listed here in the reference section and still not know, but you'll be certain to have an opinion.

1. Hans Freudenthal. "Norbert Wiener" in *Dictionary of Scientific Biography*. New York: Charles Scribner's Sons, 1976.

References:

Brittain, James E. "Introduction of the Loading Coil: George A Campbell and Michael I. Pupin." *Technology and Culture* v11, 1970, pp. 36-57, discussion pp. 596-603.

Nahin, Paul. *Oliver Heaviside: Sage in Solitude*. New York: IEEE, 1988.

Pupin, Michael. *From Immigrant to Inventor*. New York: Scribners, 1923.

Wiener, Norbert. *Invention: The Care and Feeding of Ideas*. Cambridge, MA: MIT Press, 1994.

Wiener, Norbert. *The Tempter*. New York: Random House, 1959.

Tesla: Man Out of Time
Margaret Cheney, Dell Publishing, 393 pages, 1998.
OTB, v44 n2, May 2003, pp. 58-9.

This is the second of three full-length Tesla biographies. Tesla's first major biographer was his friend John J. O'Neill, whose *Prodigal Genius: The Life of Nikola Tesla* appeared in 1944, the year after Tesla's death. Thirty-five years later Cheney had fuller access to primary and secondary sources for her biography. See also Marc Seifer's *Wizard: The Life and Times of Nikola Tesla.*

One of the many mysteries of Tesla's life is the whereabouts of some of his scientific papers, which initially went to the U.S. Office of Alien Property. In a postscript Cheney writes that "a substantial classified Tesla file" wound up in a defense research agency, but tantalizingly she does not identify the agency.

It seems one cannot write about Tesla in depth without becoming involved in mysteries. For example, Cheney repeats the story of Tesla's experimentation with his "telegeodynamic oscillator," a small engine Tesla said could be carried in his pocket and which operated via compressed air. Tesla claimed great destructive possibilities for the device through the use of resonance. "So powerful are [its] effects," Tesla said, "that I could now go over to the Empire State Building and reduce it to a tangled mass of wreckage in a very short time" (O'Neill, page 165).

Is this claim realistic? If so, the demolition industry should be informed. Why tolerate the expense, labor and danger of using high explosives when one could merely "attach a small mechanical oscillator" to the appropriate I-beam and wait a few minutes?

Cheney's and Seifer's biographies are modern, first-class works; the O'Neill biography was written from first-hand experience but is dated. All three are fascinating reading, due in no small part to their subject.

Tesla, Master of Lightning
Robert Uth, PBS Home Video, 90 minutes, 2000.
OTB, v43 n3, August 2002, p. 54.

Tesla, of course, did not master lightning. What he mastered was polyphase AC power generation, distribution and use. But try putting *that* into one short catchphrase for a subtitle.

This video documentary makes a nice companion to either of the Tesla biographies reviewed in the May 2002 *OTB*. The use of Stacy Keach's voice as the voice of Tesla adds a nice touch of Tesla's old world charm and his sense of the dramatic.

Note: the famous photograph of Tesla reading while sitting in a chair with lightning bolts surrounding him involved some legerdemain, in this case multiple exposures. Tesla was *not* sitting that close to the lightning bolts when they were photographed. Hence this picture should not appear in a documentary without explanation.

Theremin: Ether Music and Espionage
Albert Glinsky, University of Illinois, 403 pages, 2000.
OTB, v43 n3, August 2002, pp. 55-56.

For any of us who know that the frequency of an oscillator can be
changed by varying the value of a capacitor and that one "plate" of
a capacitor can be a human hand, the theremin cannot be magical or
mysterious. Nevertheless the theremin, the only musical instrument
that is played without touching it, captivated many people. These
included Lenin, Einstein (Albert and Alfred), and thousands of
others. Leon Theremin demonstrated his electronic musical
instrument for Lenin in1922; then he gave Lenin a lesson on the
instrument by guiding his hands.

   Glinsky's book saves Theremin from having the reputation of
being a one-act inventor. He invented almost constantly during his
long life (1896-1993). In 1927 he demonstrated his television system
to Stalin. The system was considerably advanced over that of
Alexanderson at General Electric. (Theremin's television had a
screen resolution of 100 lines and was capable of use in natural
daylight. In 1926 he had demonstrated a television screen five feet
square).

   Though Stalin could see no educational or entertainment value to
television, he immediately realized the spying potential of
Theremin's machine and "appropriated" it for state use. Theremin
was rewarded for his work with a government coupon for "a big food
parcel."

   Theremin spent much of the 1930s in the U.S., demonstrating his
musical instruments (and feeding corporate and industrial
intelligence back to the Soviets). Shortly after he was "recalled" to
the Soviet in 1938, he fell victim to one of Stalin's purges. He was
found guilty of "treason to the Motherland" and sentenced to eight
years' imprisonment in the infamous Kolyma camps in eastern
Siberia. Many people sent there never returned, regardless of the
length of sentence. They died mining gold under inhuman conditions,
working in temperatures that could go to -94°.

   Theremin was put on a cattle car with other prisoners for the 5,200 mile railroad trip to Vladivostok. There the prisoners were put aboard huge "slave" ships for the trip across the treacherous Okhotsk Sea. (Actually, Theremin was lucky just to get to Kolyma. It could have been otherwise. In 1936 the ship Dzhurma, on a similar journey, got stuck in autumn ice. When it finally arrived at its port that spring, none of the 12,000 prisoners in its hold were alive.)[1]

Always the innovator, Theremin designed a gold mining process that improved the one used at Kolyma. Fate soon intervened and he was recalled again to Moscow, there to help in the design of airplanes for use in WW II. (Evidently the fact that Theremin knew little about airplane design was counterbalanced by Russia's great need for new airplanes). Later he designed a resonance cavity transmitter that was hidden in a wall plaque in Averell Harriman's study in the American ambassador's house in Moscow. Still later Theremin was to have a kind of revenge on Stalin. He designed a miniature listening device that Soviet authorities hid in Stalin's desk.

For another view of Theremin's life and music machine, there is a video documentary produced by Steven M. Martin: *Theremin: An Electronic Odyssey*, MGM/UA Home Video, 1995, DVD, 84 minutes, $19.98.

To hear Theremin music in a venue other than 1950s sci-fi movie soundtracks, start with Clara Rockmore's *Art of the Theremin*, Delos Records # 1014, $11.98. This 1992 recording, now on CD, was co-produced by Robert Moog, of synthesizer fame. Of course, everyone's heard the theremin-like passages in the famous Beach Boys' song "Good Vibrations."

Glinsky's top-notch book got deserved rave reviews and won the ASCAP-Deems Taylor Award in 2001. The book, the video and the CD are all available through *amazon.com*. And if you really get bitten by the theremin bug, you can buy one, assembled or kit form, at Moog's website: *www.bigbriar.com*.

1. A newer publication casts doubt on this story. See Martin Bollinger, *Stalin's Slave Ships*, Praeger Publishers, 2003.

3 Strikes Camp Stories
Karl Laurin, *www.lulu.com*, 14 pages, 2008. Illustrations by the
   author.
Pittsburgh Oscillator, v23 n1, March 2008, p. 8; reprinted in *AWA
Journal*, v49 n2, April 2008, p. 23.

These three short short stories, gathered here for the first time,
originally were published in the *Pittsburgh Oscillator*. When seen
and read together they classically become something "larger than
their sum."

I liked the stories for two reasons. First, they contain no fluff. Each
word in them has been carefully chosen to produce the exact effect.
Second is the effect itself. The stories adroitly capture the ambiance
of a small mining enclave high in the Sierra Nevada Mountains in the
early 1920s, and the effect on the camp of Cookie's one-tube radio.
In a pinch, the radio is modified into a transmitter when "Romance
calls."

Laurin acknowledges his debt to and appreciation of Bret Harte's
Roaring Camp in the preface to the stories.

Look! One can almost see Tonto and the Lone Ranger riding into
3 Strikes Camp on a bright summer day–humidity, oh, about 10%–
in pursuit of a small band of desperados. There they listen carefully
to the stories of Cookie, Kitz and especially the "Perfesser," who had
"taught geology and metallurgy at Stanford but was kicked out for
coed trouble". Then, with new clues, they refill their canteens, saddle
up, and descend the steep mountain trail to the broad plain stretching
into the sunset below, eventually, of course, establishing Swift
Justice for a Grateful Nation.

Tickling the Crystal: Domestic British Crystal Sets of the 1920s
Ian Sanders and Carl Glover, Bentomel Publications, vol.1, 256
    pages, 2001; vol. 2, 208 pages, 2004; vol. 3, 240 pages, 2005.
Vol 1: *OTB*, v43 n1, February 2002, p. 63.
Vol 2: *OTB*, v45 n3, July 2004, p. 14.
Vol 3: *AWA Journal*, v47 n2, April 2006, p. 25.

Volume 1: Many of us had a crystal set, toy or otherwise, as a
child, but be warned–this book may well produce a major change in
the lives of those who approach it thinking they are no longer
interested in such things.

The glossy, heavy paper used in the book for benefit of the
photography gives it a heft that hints at a corresponding quality of
content. In this readers and viewers will not be disappointed.

The first thing noticed is the high quality of the photographs,
beginning with the striking picture of a Brownie No. 2 crystal set
with tube amplifier on the front cover. Those contemplating writing
their own illustrated book would do well to note the equipment and
programs Glover used to produce the virtually flawless photos here.
About 200 are included, some in color.

The text is on a par with the photos. Sanders obviously has wide
experience with British crystal sets and their history. He put ten years
into the research for this book, and that work is evident on every
page. He devotes a chapter each to the crystal set era (about 1922 to
1928), General Post-Office registration, manufacturers, circuits,
detectors, aerials and earths, headphones, cabinets, novelty sets, note
magnifiers (amplifiers), hybrid (crystal/tube) receivers and the home
construction of sets.

The bulk of the book consists of chapter 13, "Pictorial Dictionary
of Crystal Sets." The description accompanying each illustrated set
details manufacturer, model, year, construction, tuning and detector
types, original price and other information.

Occasionally some interpretation from the British idiom is
necessary. A prominently-featured 1926 quote by BBC Chief
Engineer P. P. Eckersley advises crystal set users whose reception is
unacceptable:

"get a valve set or double your aerial, or look to your earth . . . ." The American reader will translate this to: "get a tube set or raise or lengthen your long-wire, or check your ground connection . . . ." An interesting note about antennas I've seen nowhere else: it is suggested that the height of a long-wire should not exceed 60 feet, "since this would require too much of the total length being consumed by the down-lead portion."

About 20 pages of appendixes include a directory of manufacturers (complete with original addresses!), GPO registration numbers, brands of crystals, a list of crystals and minerals, note magnifiers, headphone models, and circuit diagrams. Two indexes–general and model name/number–are included, another indication of the thought and work that went into the production of this volume. As anyone who has ever prepared an index for a book knows, the work is tedious and time-consuming, and the publisher wants the copy *right now*.

From cover to cover, this book is first class.

Volume 2: Volume two of Sanders and Glover is obviously done with the same loving care and attention to detail as the first.

"A very good imitation of silence can be maintained in the wireless den." This sentence, from a 1924 letter to *Modern Wireless* (London) quoted in the book, is, I think, about as British as can be. So is the book, and we commend Glover again for his flawless photographs and Sanders for his meticulously detailed descriptions of British crystal sets, crystal wavemeters and "note magnifiers" (amplifiers).

Sanders says, ". . . new material on the subject continues to accumulate and the possibility of a third volume is under consideration." We say, "Bring it on."

Volume 3: In the Victoria and Albert Museum a few years ago I saw a 1920s British crystal set that made my mouth water. I remember distinctly the over-engineered gleaming solid brass thumbscrews, each the diameter of a quarter, on the "aerial" and "earth" terminals. I briefly considered breaking the display glass, grabbing the set and making a dash for it, but I was able to curb my enthusiasm.

Now we have the third volume of the Sanders and Glover set and I am able to slake my thirst while remaining within the law. Beautifully engineered and cheaply engineered crystal sets were made in the U.S. and in Britain, and both types of British sets are shown here.

Have Sanders and Glover now documented all British crystal sets? Not likely, but, as Sanders states, more ephemera appears in volume three than in the earlier two. Ephemera it may be, but fascinating it is. So we have, scattered among the gorgeous photographs, contemporary cartoons, readers' comments and questions, directions on how to wash and "revive" crystals and, as the saying goes, much, much more.

The cost (£29.95) of the book is certainly worth it, even if only to get your own copy of the 1920s picture on page 4 showing "James Roberts listening to radio reports on the King's health on his crystal set in Epping, Essex" or the picture of the "largest crystal set ever made" on page 85.

This now-classic set is highly recommended.

Note: the following volumes have been added to this series:

Volume 4: BVWS Books, 2008, 280 pages, hardback.
Volume 5: BVWS Books, 2010, 252 pages, hardback.
Volume 6: BVWS Books, 2012, 80 pages, softback.

Tube Guys
Norman H. Pond and others, Russ Cochran, Publisher, 447 pages, 2008.
AWA Journal, v50 n1, January 2009, pp. 35-36.

In chapter eight of this book, contributing author Bob Phillips, inventor of the ubitron, describes his early work on the device: "I programmed the ballistic equations of motion for a magnetically undulated electron beam interacting with a TE_{01} mode in an unloaded waveguide on an analog computer . . . ." Okay, so we've obviously left behind the prosaic world of 01A's and 6AU6's here and entered the esoteric world of klystrons, cavity magnetrons, traveling wave tubes, backward wave oscillators, gyrotrons, ubitrons, and linear accelerators. Why was it necessary–or even desirable–for us to do so?

That question is answered in chapter two, "Protecting Britain." World War I had shown that Britain was vulnerable to bombing attacks from enemy aircraft. Something had to be done. Of 53 ideas advanced for British air defense all were found wanting in effectiveness. But British gadfly and overall smartie Robert Watson-Watt became excited when he learned that airplanes produced disturbances in radio waves when they flew between transmitter and receiver antennas. Watson-Watt wrote a few well-placed and well-timed memos to higher-ups in the government and military (he knew *many* VIPs). As a result the British government moved quickly to form the "Radio Echo Detection System," now known as radar. (For more about Watson-Watt, who was knighted in 1942 for his wartime efforts, see p.129).

From the start radar wanted high frequencies at high power for high accuracy. What was needed were devices in the one to 100 GHz range that could handle power in the thousands of watts. Even today solid state devices can handle only hundreds of watts in the microwave range; the high power/high frequency combination demanded by radar requires vacuum tubes. The men who "stayed with tubes" after the transistor revolution and invented and developed the devices named above are called "tube guys," hence the book's title. The first part of the book devotes historical chapters to each of these devices.

Microwave tubes come in many sizes and shapes, from the funny-looking black tube with the one pin too long occasionally seen at flea markets to the monster supported by steel chains that is partially pictured on the book's dust jacket. Of course all AWAers own at least one of these devices, buried deep inside their kitchen's "radar range." Magnetrons for use in microwave ovens required much money and many man-hours to develop; Pond tells us that they are now produced at $7 per unit.

The second part of the book comprises single-chapter histories of about two dozen companies producing microwave products aimed at radar, communications, radiometry, microwave ovens, industrial heating and drying, etc.

Little in this book will interest the average weekend flea market radio hunter, but much in it will be of interest to "tube guys."

Tube Testers and Classic Electronic Test Gear
Alan Douglas, Sonoran Publishing, 166 pages, 2000.
Pittsburgh Oscillator, v15 n4, December 2000, p. 7.

A truism has developed in the antique radio world–Alan Douglas pays attention to detail. And his readers are the beneficiaries. The latest demonstration of this truism is Douglas's new book, devoted largely to tube testers but also covering VOMs, VTVMs, Q meters, grid-dip meters, bridges, signal generators and tracers, and oscilloscopes.

The organization of Douglas's new book parallels that of his well-known series, *Radio Manufacturers of the 1920's*. An information-packed introductory chapter is followed by an alphabetical presentation of manufacturers, with further information pertinent to each.

The introduction for tube testers clearly explains the differences between emission, dynamic and mutual-conductance testers, and evaluates each method. Douglas describes various modifications and calibrations that can make tube testers more accurate, and finishes the introductory section with the essay "When Tube Testers Disagree" (which they often do).

Have you ever come away from a tube tester a little more confused than when you approached it? Several reasons are projected to explain this phenomenon. One of these reasons I did not want to think likely, but Douglas spells it out clearly in his chapter on Eico tube testers (and this explanation is a good example of the attention paid to detail in the book). It seems that Eico's roll chart data is sometimes less than accurate! The roll chart for the well-known Eico 666 went through nine editions, with some editions listing switch settings that contradicted the settings in other editions. Douglas states, "Basically, it took Eico nine tries to get it right!"

Sometimes troubleshooting equipment is used to troubleshoot other troubleshooting equipment, and Douglas details several cases of how inoperable test equipment picked up at a flea market finally became useful additions to his workbench. (In this regard I sometimes suspect that the world may in fact have too many oscilloscopes).

Douglas has kept his book from being a static compilation of technical facts by scattering human interest stories throughout. "Origins of the D'Arsonval Meter" and "Why Hewlett-Packard, not General Radio, is known for R-C oscillators" are examples. His wide knowledge of company histories also allows him occasionally to characterize in terms only a long-time radioman would understand, as in the phrase "General Radio (GenRad) now has more of a '"they're still there' reputation . . . ."

More detail: Douglas gives Samuel Hunter Christie his due as the inventor of the Wheatstone Bridge (this credit is rarely seen in radio history literature), and he includes the nugget that Allen B. Du Mont sold his patent for the "Eye Tube" indicator to RCA for $20,000.

As with *Radio Manufacturers* volumes, Douglas greatly enhances his book with the inclusion of many black-and-white photographs and reproductions of original ads from various magazines and manufacturers' catalogs. The 166-page book is indexed, and the back cover features a photograph of the author pretty much surrounded by what it is tempting to assume is some of his favorite test gear.

If you like using, or just looking at and being around, nice old electronic test gear, you will like this book. There is another truism–this one from the realm of literary criticism–which says that good literature either instructs or entertains. Douglas's book does both.

Uncle Tungsten: Memories of a Chemical Boyhood
Oliver Sacks, Vintage, 337 pages, 2001.
OTB, v44 n1, February 2003, p. 33.

Occasionally we notice books outside AWA's purview but still of potential interest to readers of *The OTB*. Such is the case with Sacks's delightful autobiography of his childhood and adolescent years.

To say Sacks had an unconventional childhood is an understatement. He was largely self-taught with the help of books, museums and a variety of brilliant aunts and uncles who fostered his scientific interests and never discouraged his early investigations and experiments. His Uncle David, a.k.a. Uncle Tungsten, operated a tungsten-filament light bulb factory and inspired in Sacks a love of that metal, and metals in general.

While other students his age were puzzling over the periodic table of the elements, Sacks memorized it and investigated its history, from before 1869 when Mendeleev devised his version of it to the definitive "tweaking" of it in 1913-14 by Harry Moseley. (Moseley was killed in WW I at 28).

Other outstanding experiences of Sacks's formative years included repeated beatings by a sadistic headmaster in a boarding school where Sacks had been sent, ironically, for protection from the bombing of London during WW II. It was always understood by Sacks and his parents, both of whom were doctors, that he, too, would be a physician. To that end his mother arranged for the 14-year-old Sacks to perform a human dissection. The cadaver in question, appropriately or otherwise, was that of a 14-year-old girl.

What happens to a person who has had such a childhood? Today Sacks is a distinguished neurologist living in New York City. His many books include this national bestseller.

Victorian Internet: The Remarkable Story of the Telegraph and the Nineteenth Century's Online Pioneers
Tom Standage, Orion Books Ltd., 216 pages, 1999.
OTB, v45 n3, July 2004, p. 14.

Likening the telegraph to the internet is in some ways a stretch. In the Victorian world telegraph messages generally went to just one recipient; today's spam was unknown. Few people owned their own equipment to access the telegraph system; today millions own computers that access the internet. Video and audio features were unknown to telegraph users, but they are a big part of the internet. But Standage's popular account of wire telegraphy is so engaging that the reader doesn't care about the differences between the telegraph and the internet.

Of course, there were some obvious parallels. The telegraph and internet alike were initially resisted by the public and thought of as toys. Then both systems were so thoroughly embraced by society that information overload became common. Victorian essayist Lytton Strachey described the telegraph as "that distressingly useful invention." Victorian architect William Morris declared that "for their brutalizing influence upon humanity telegraphs were as much to be blamed as were railways."[1]

Main telegraph lines in England became so clogged with messages that they were replaced with pneumatic tubes carrying messages on paper. Hence the recipient of a telegram in some cases received the actual paper form that the sender wrote upon.

Another parallel was the initial idealistic hope that each system would promote world peace. Standage says, "Given its potential to change the world, the telegraph was soon being hailed as a means of solving the world's problems." But neither the telegraph nor the internet promoted peace, probably because both systems were used by human beings.

This book makes educational and entertaining reading.

1. Rollo Appleyard, *Pioneers of Electrical Communication.* Macmillan, 1930, page 255.

Vintage Radio Alignment: What It Is And How To Do It
Bret Menassa, DVD edition, 52 minutes, 2009.
AWA Journal, v51 n2, April 2010, p. 58.

Menassa adds a sense of immediacy to his latest instructional DVD by presenting his talk before a live audience. This required the usual prep work selecting radios, but unlike his earlier DVDs, this one was "one-take" and includes comments and questions by members of his audience.

He cleverly begins his talk by explaining and demonstrating the action of his guitar as an instrument for transferring power. The importance of tuning and resonance in the guitar are used as a metaphor for those same characteristics in radio reception.

His first tuning demonstration is for simple tuned radio-frequency sets (or TRFs, an acronym that should have been explained), then shows the difference between aligning TRFs and superhet sets. He then demonstrates in great detail the alignment of All American Fives (including those with six tubes), shortwave and multi-band radios, and FM receivers. The similarities between AM and FM alignment are presented.

Most of Menassa's alignment set-ups use RF induction into the radio's antenna, with a digital AC meter across the speaker voice coil as an output indicator. Most alignment is done without reference to manufacturers' instructions, but Menassa indicates instances when manufacturers' procedures should be followed, as when, for example, sets have wave traps in antenna circuits. Throughout the presentation, he mentions things–such as keeping generator output low, or the delicacy of iron-core inductor adjustments in FM receivers–that could be known thoroughly only by someone who has been there and done that many times.

(See also *Antique Radio Restoration*, page 7).

Vintage Radio Identification Sketch-Books
D. H. Moore, Olde Tyme Radio Company, 1954-1984?
Pittsburgh Oscillator, v1 n3, December 1986, p. 5.

Where do you go when it isn't in Sams, Rider's, Beitmann, the RCA manuals or Gernsback's encyclopedia? Certainly one place to look is the *Vintage Radio Identification Sketch-Books* of D. H. Moore.

The antique radio hobby owes a great debt to Moore for producing these books. Together they represent many hundreds of hours of thought and work. When Moore was unable to find a schematic for an "off-brand" radio in the 1920s and '30s, he solved his problem by drawing one of his own. These often included such details as winding specifications for coils. But he didn't stop there. He also made hundreds of sketches showing thousands of details of these sets.

Interspersed with these schematics and graphic sketches are Moore's textual sketches, in which he tries, as he says, to capture the background rather than the history of early radio. This attempt leads to such essays as "What's In A Name?", in which Moore lists the following names from early radio media: Reber Boult, Merwin Dobyus, Hjalmar Stromberg, Valeski Bari and Toufic Moubaid. Dozens of other oddities abound, such as the photograph of the 1922 Lady's Radio Garter in action.

Moore's thoughts range pretty widely in these essays. Who, for example, expects to find comments about gays and women's lib in vintage radio literature, or frequent references to the general decline of Western civilization, from the peak it apparently enjoyed during Moore's (and radio's) formative years to the present-day level?

Schematics, sketches and essays make up about 1100 pages in all, and these nine volumes will give hours of pleasant reading and can yield literally thousands of facts and details found nowhere else. See, for example, on page 109 of Volume One the "Bibliography of Forgotten Circuitry," an index to hundreds of circuits appearing in early radio magazines.

The books are not without error, however, and right at the top of the list of these are the innumerable typographical and printing errors that appear throughout. It is a shame that this valuable work should be marred by mistakes that were so easily correctable.

These somewhat strange, opinionated, fascinating and useful volumes may be ordered from Olde Tyme Radio Co., Silver Spring, Maryland.

I: *Introductory Essays*, 131 pages.
IIA: *Broadcast Receiver Schematics, A-L*, 151 pages.
IIB: *Broadcast Receiver Schematics, M-Z*, 140 pages.
III: *Short-Wave Receivers*, 152 pages.
IV: *Power Supplies & Amplifiers*, 136 pages.
V: *Hardware and Accessories*, 114 pages.
VI: *Test Equipment Schematics*, 72 pages.
VII: *General Technical Data*, 100 pages.
VIII: *Earlier Years: 1910-1924*, 164 pages.
IX: *Master Index.*, 26 pages.

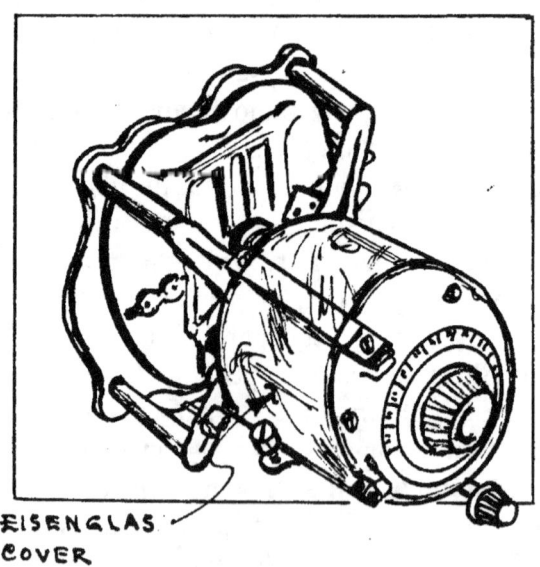

EISENGLAS
COVER

Fig. 1. Detail of Acme D-Coil, from D. H. Moore, volume II.

Vintage Radio Redux: Karl Laurin in the Pittsburgh Oscillator
Karl Laurin, Pittsburgh Antique Radio Society Publication 14,
www.lulu.com, 110 pages, 2008.
AWA Journal, v49 n3, July 2008, pp. 24-25.

In June 1992 Pittsburgh Antique Radio Society member William
J. Johnson, Jr. sent a letter to the *Pittsburgh Oscillator* explaining
why he was dropping his membership in the club. He wrote, in part:

> *As long as antique radio is asserted toward the professional
> level it is condescendant to the junk that constitutes the
> delight of the majority of its own followers . . . . The only
> personal integrity remaining for the casual radio devotee is
> limited to a very elementary past technology, and to his own
> two hands in manual activity. I have found the Pittsburgh
> Oscillator to be outside that concept.*

At the time, I could not understand these objections. But now that
I've seen Laurin's articles collected as a book I begin to appreciate
Johnson's concerns. I think he would have liked this book.

With this publication the PARS series breaks away from the
"professional level" of radio (though Laurin's book *is* written in a
professional manner) to engage the expression of amateur
electronics' nitty-gritty side. Three examples will suffice. Why spend
big bucks at the Shack for a large heat sink? Instead, Karl I-never-
throw-anything-away Laurin simply uses part of an old aluminum
storm door frame. Or suppose you don't have a switch for the battery
in your latest transistor project. No problem–just twist two bare wires
together (*on*) or untwist them *(off)*. On page 13 Laurin shows us his
home-brew active antenna. The chassis for his project? A toilet paper
tube.

But it's not just about money-saving hints and kinks. Laurin also
provides the reader with technical insights into why some circuits
(tube or transistor) don't work and why others work the way they do.
See, for example, his diagnostic techniques in fixing a tiny,
oscillating Japanese pocket radio, or his thorough explanation of how

189

to tune a 1920's Paragon one-tube radio. The owners of that Paragon RD-5 were lucky to find Karl when they brought their grandfather's radio to a PARS service clinic.

Most of this stuff will not be found in the average introductory textbook; knowledge of it was gained from years of practical experience in radio and television service work and from teaching electronics at a tech school. As the age-old advice goes, to know a subject well, teach it.

This how-to and personal-experience collection of off-the-beaten-path essays is happily recommended. Photographs and hand-drawn schematics enhance the text. Several of the author's cartoons and drawings are included. It's at once informative and fun to read.

What Hath God Wrought: The Transformation of America, 1815-1848
Daniel Walker Howe, Oxford University Press, 904 pages, 2007.
AWA Journal, v49 n2, April 2008, pp. 23-24.

What Hath God Wrought–wouldn't that make a great title for a history of telegraphy? Yes, but here Howe uses it as a title for a dense, thorough, highly-scholarly description of a 33-year period of American history. He uses Morse's famous phrase (taken from the Bible) because he believes that the so-called Jacksonian period was largely wrought by a communications revolution.

Explication of the social, military and political effects of communication abound in the book, but the telegraph per se receives a treatment of only a dozen-or-so pages. Nevertheless, these include many facts and insights. For example, in 1846 only 146 miles of telegraph line existed. The war with Mexico so stimulated telegraph growth that by 1848 the lines had reached New Orleans. By 1850, 10,000 miles of telegraph wire had been installed in the U.S. Morse's telegraph worked so well that it gradually replaced the earlier Wheatstone telegraph in England. (Morse's telegraph recorded; Wheatstone's didn't). Initially the Russian tsar refused to install telegraph lines, fearing the device would "facilitate political opposition." And–hard to believe now–at first the railroads didn't recognize the importance of the telegraph for their operations, particularly scheduling, and even more important, the avoidance of collisions.

Nor was the tsar the only one unhappy with the telegraph revolution. Howe includes the wonderful quote by Henry David Thoreau: "We are in great haste to construct a magnetic telegraph from Maine to Texas, but Maine and Texas, it may be, have nothing important to communicate." Here of course Thoreau, the man of letters, was just plain wrong.

When Radio Was Young: Questions and Answers About Early Pittsburgh Radio
Bill Beal, Alice Sapienza-Donnelly, Rick Harris, Jr., Wilkinsburg
    Commission, Inc., 1995.
Pittsburgh Oscillator, v10 n4, December 1995, p. 4.

  Nostalgia drips from this book. All PARS members–and many
others–will want to own a copy. Not only is it chock full of radio
history, it's chock full of radio history as it was experienced by
us–millions of 'Burghers, Pennsylvanians, and indeed, millions of
people throughout the world.
  The book is dedicated to Frank Conrad and carries a report about
the Conrad Project. A Bill Hofscher portrait of Conrad is
prominently featured in a montage painting reproduced on the back
cover; there are also many black-and-white photographs used
throughout, some borrowed from the collections of the people
featured in the book. And veteran Pittsburgh animation artist Lee
Hartman has a "character study" of the first five Pittsburgh radio
stations on the front cover.
  The book is organized around three main chapters, each dedicated
to a different decade in radio history, beginning with the Twenties.
An introductory essay at the beginning of each chapter describes
radio's cultural setting and influence. The remainder of each chapter
is in question-and-answer format. Some of the questions the book
asks (and answers) are:

>   *What future President unknowingly made his radio debut*
>   *on January 15, 1921 in Pittsburgh?*
>   *What's the story behind the broadcast of a human heart*
>   *beat back in 1924?*
>   *What was the name of the early soap opera that originated*
>   *in Pittsburgh?*
>   *Where did Rosey Rowswell begin announcing Pirate*
>   *broadcasts?*
>   *What radio engineer made it possible for motorists to*
>   *receive radio while going through the Liberty Tubes?*

Among the answers one finds a panorama of people who were part of Pittsburgh radio history: Marconi, Fessenden, Sebastian Sapienza, H. P. Davis, Brian McDonald, Ted Yearsley, Slim Bryant, Ed Schaughency, Bill Brandt, Paul Shannon, Art Farrar, John Scigliano, Davey Tyson, Jack "E. Z. Credit" Logan, Joseph Baudino, Father James Cox, Lawrence Welk, Ed and Wendy King, Bill Hinds, and on and on.

The over-large type used might give the book a less-than-professional appearance to some, but will be welcomed by the many hyperopic among us. And we would like to have seen at least a personal names index so that information on the many radio personalities included could be found quickly.

But these minor flaws won't prevent any reader from spending a pleasant evening with this book, especially if a Westinghouse RA/DA is softly playing KDKA in the background.

Where Discovery Sparks Imagination: A Pictorial History of Radio and Electricity
John D. Jenkins et al., American Museum of Radio and
    Electricity, 218 pages, 2009.
AWA Journal, v50 n3, July 2009, p. 67.

The keyword in the title is *Imagination*. With this book Jenkins wished to recreate the "sense of magic and discovery" to be found in the history of electricity and radio. In this he has succeeded, helped in no small way by hundreds of beautiful color photographs of electrical equipment, beginning with an 18th century English lodestone and continuing through to a "moon" tube of the late 1960s.

The items chosen to be photographed for this book, and the photographs themselves, are so attractive that no matter how much one wishes to read the book straight through, it is too inviting to flip it open at random and be captivated by whatever appears. Examples: striking photos of gold leaf Leyden jars, a Volta pistol and cannon, Hertzian wave apparatus, a cobalt blue Vic's Vapo-Lite, an Edison "chemical" power meter, any number of early electric motors and induction coils, telegraph equipment and telephones–all that before Jenkins gets to radio in the second half of the book.

Here the photography is as appealing as in the first half. Seeing the photograph on page 150 of an Atwater Kent Model 5 has to be almost on a par with seeing the actual thing itself. A similar thing can be said for the close-up photograph on page 168 of a Zenith split dial. Talk about capturing a "sense of magic and discovery"! Congratulations to Jenkins and Sam Spencer for their stunning camera work.

The accompanying text was clearly written by someone with much experience in the subject. It points out many details that most readers will not have been aware of. For example, I've had an RCA Radiola Super VIII in my living room for many years, but it wasn't until I read the caption for the photograph of one in the book that I learned about the "candlesticks" designed into the speaker cover. I had to go look for myself. Yep, they're there. Their presence may have gone unnoticed by other VIII owners. Might they be merely a matter of visual imagination?

So I guess I now have to come up with some comment to keep this review from being a total rave. Right at the top of my very short list here is the tiny print used in some of the photo captions. It's even worse, to my eye, when the letters are white against a black background. I understand the motivation–the authors want to give as much information as possible about each photo, but the heavy, glossy paper stock used throughout the book to enhance the photos cannot be cheap. So, small font. End of criticism. I hope all my readers will sooner or later run into a copy of this beautiful book.

Wireless: From Marconi's Black-box to the Audion
Sungook Hong, MIT Press, 248 pages, 2001.
OTB, v44 n1, February 2003, p. 34.

This book represents scholarship at its best. Not only does Hong have a keen eye for the details he locates in his thorough research, but he couples that with an acute ability to see and analyze the meaning of the facts he finds.

When he reports facts, they are always carefully documented (the 200 pages of text are followed by 29 pages of notes). When Hong speculates as to the meaning of his facts he always carefully lists the reasons for drawing the conclusions that he does.

Hong's essays discuss how emphasis on Hertzian optics tended to delay radio's development, the Marconi *v.* Lodge struggle, Marconi's treatment of Fleming and vice versa, and the effect that Maskelyne's "jamming" of Marconi's demonstration broadcast had on Fleming's career. There is also a very detailed analysis of the long and tortuous road from Edison's "Effect" to de Forest's triode.

The book will give each of its readers new insights into why radio history evolved as it did and why the various players did what they did. It is highly recommended.

Wireless Radio
Lewis Coe, McFarland & Company, Inc., 204 pages, 1996.
Pittsburgh Oscillator, v12 n2, June 1997, p. 7. Reprinted in *OTB*,
v42 n3, August 2001, pp. 31-32.

The usual chapters about KDKA, crystal sets, and antique radio collecting are here, but the real value of this slender book lies in the extensive survey the author has made of all major uses of radio. The survey ranges across "the Vast Continent" (William Crookes's phrase for the radio spectrum) and from the life of James Clerk Maxwell into the 1990s. So we also have chapters on marine radio (the first practical use of radio), amateur radio, point to point, military radio, radar, police radio, television, cellular and satellite telephones, portable radio, the attempt to transmit power via radio, etc.

Here are clear, though necessarily brief, descriptions of Ampex video recorders, the DEW line, diathermy, ELTs, Globalstar, GMDSS, INMARSAT, Loral, LORAN, microwave ovens, MSAT, NAA, OMEGA, Qualcomm Sincgars radio, Very Long Array, etc. Pitcher Nolan Ryan even gets into the act in a discussion of the use of radar in sports.

Most readers will probably recognize at least some of these topics. Coe, born in 1911, points out that he lived through the development of most of them and he took many of the photographs in the book.

Do you think you know what's happening in radio? Consider this: a Boeing E6A or Lockheed EC-1300 takes off carrying with it a coil of wire weighing almost a ton. Aloft, the wire is trailed out behind the plane for about five miles. "When the aircraft supporting the wire flies in a tight circle, the wire tends to assume a near vertical position that is essential for best results." This wire is used as a transmitting antenna, sending signals to submerged submarines using Very Low Frequencies (VLF). Other land-based antennas over 50 miles long transmit to submarines using frequencies as low as 30 Hz, with a range up to 5,000 miles.

No book is perfect, and Coe may be content to know that great law includes even this highly readable book. Many AWA members know that KDKA pioneer Conrad's first name was Frank, not Charles.

Coe has also written *The Telephone and Its Several Inventors: A History* (McFarland, 1995) and *The Telegraph: A History of Morse's Invention and its Predecessors in the United States* (McFarland, 1993).

Wizard: The Life and Times of Nicola Tesla
Marc J. Seifer, Citadel Press, Carol Publishing Group, 540 pages,
    1996.
OTB, v43 n2, May 2002, p. 52.

Seifer's scholarly study is based on his Ph.D. dissertation and is
heavily documented. Lomas's book, *The Man Who Invented the
Twentieth Century,* is a short, popular account with no footnotes (see
page 106). Both books present a life that can only be described as
fascinating.

Poor Tesla. Nothing in his life was ordinary. He was by turns
supremely brilliant and naive, wealthy and destitute, legally and
illegally robbed, and a visionary genius who sometimes earned his
keep by digging ditches. He never married and apparently formed
few close relationships. He lived in hotels for most of his adult life,
and often shared his rooms with pigeons. He learned through an
article on the front page of the November 6, 1915 *New York Times*
that he was to share a Nobel Prize with Thomas Edison. In a
subsequent interview he expressed his gratitude to the world for
finally recognizing the value of his work, only to learn later that the
Times story was false.

Tesla (pronounced TESH la) gets an A+ in electricity, but a much
lower grade in business acumen. Legend says he sold the 40 patents
for his AC polyphase system to Westinghouse for $1,000,000 (some
writers report a much lower figure), plus a royalty of "one dollar per
horse-power." Later he either relinquished this royalty, surely worth
millions, or sold it back to Westinghouse for a paltry sum. Still later
he testified that he thought at the time that Westinghouse was to have
the royalty back only temporarily. For much more detail on these
shenanigans, see the Seifer book.

One of the reasons Tesla's life is so intriguing is that so much of it,
including the exact date of his death, ends in mystery. Some of this
mystery is attributable to what Seifer says was Tesla's "irritating
habit of . . . seeing projects complete before they actually
materialized . . . ."

Surely Tesla's most important idea, i.e., the idea that had the most potential for benefitting humanity, was the transmission of electrical power without wires. Tesla sold his patents for this system to J. Pierpont Morgan. Morgan "sat" on these patents and thus stymied further development of them, possibly to protect his own considerable financial investment in Tesla's earlier wired system of AC power transmission.

So here is the question: could Tesla's system of wireless power transmission be made practical? If so, why has no one done it? Patents for the system were published 100 years ago. The fact that no one has achieved widespread wireless transmission of power to date would seem to indicate that Tesla's scheme cannot work.

Readers may argue forever over Lomas's list of major inventions claimed for Tesla: radio, the hydroelectric generator, AC power, fluorescent light, the rotary engine, the bladeless turbine and radio imaging. But who could quibble with an author who dutifully reports one of Tesla's lesser-known inventions, a significant enhancer of peristalsis? Tesla used the device, a variable-frequency vibrating platform upon which one stood, to successfully treat his indisposed friend Mark Twain.

These books can't help but make many of their readers Tesla simpaticos.

Words at War: World War II Era Radio Drama and the Postwar Broadcasting Industry Blacklist
Howard Blue, Scarecrow Press, Inc., 407 pages, 2002.
OTB, v44 n2, May 2003, p. 58.

This book follows a dozen or so lives in radio through two distinct phases–the heyday of polemical writing and radio drama between 1936 and 1945 and the distinctly-soured postwar radio atmosphere in the days of McCarthyism.

In attempting to explain how the war of words originated, the author provides a recipe, literally, which includes such ingredients as the advent of radio, the radio drama concept, FDR's New Deal propaganda broadcasts, Orson Welles's broadcast of "The War of the Worlds," and, of course, World War II. Combined, these factors and others produced a vigorous war of words directed against fascism, Nazism, and Japanese war efforts. The author states: "In 1940, more than 10,000 hours of drama were broadcast over the air. This represented more than 30 percent of the available air time."

But soon after the war ended, in an ironic reversal of fortune, the war of words was turned against these same idealistic drama writers and producers. Their liberal views in defense of democracy and human rights went out of fashion in an American society now thought to be riddled with Communism at all levels. The shenanigans of the McCarthyites revealed in the last two chapters are difficult to fathom, particularly after the civil rights protests and advances of the '60s and '70s.

The book is extensively researched and documented, with notes, bibliography and index.

Addenda

DECORADIO: the most beautiful radios ever made
Peter Sheridan, Schiffer Publishing, Ltd., 352 pages, 2014.
Pittsburgh Oscillator, v30 n2, June 2015, p.10.

Justified rave reviews are appearing about this book.

Sheridan calls these the most beautiful radios ever made. The phrase reminds me that Sheridan's book itself is in the running for the prize of the most beautiful radio picture book ever printed. Other candidates in this category are John D. Jenkins's *Where Discovery Sparks Imagination* and Eric Wenaas's *Radiola*.

Sheridan's book includes about 380 "living color" photographs of radios from the '30's, '40's and '50's. The scope is international, with material from 15 countries.

**1939 Sparton 500C. Designer: Walter Dorwin Teague.
From the dust jacket.**

The glossy black pages used throughout the book have been criticized for obscuring some of the details of the radios. This is true; however, it is a small price to pay for the overall impact achieved.

The beauty of the photos creates the danger that the text of the book will be overlooked. This is unfortunate. Sheridan's earlier book, *Radio Days–Australian Bakelite Radios,* puts him in good stead for writing the present volume, and his essays are full of insights. Examples: he states that Deco radios are seldom used as exemplars of the Art Deco style in general. One wonders why. He also tells us that although we see these radios as "chic" today, that was not the case when they originally appeared. They were often thought of as "bottom of the line" in a manufacturer's list of products.

Do not let the price ($79.99) dissuade you from *somehow* experiencing this visually luscious book. Note: I am not hereby advocating theft; if you must, you can borrow my copy (for a few minutes).

Majestic 5T-0, page 239.

Tesla: Inventor of the Electrical Age
W. Bernard Carlson. Princeton University Press, 500 pages, 2013.
Pittsburgh Oscillator, v31 n3, September 2016, p. 14.

Seventy years after his death, Nikola Tesla still excites us, and will continue to do so for many more years.

Carlson's full-length biography now takes its place beside those of Martin, *Invention, Researches, and Writing of Nikola Tesla* (1894), O'Neill, *Prodigal Genius* (1943), Cheney, *Tesla* (1986 and 1999), and Seifer, *Wizard* (1996).

So, was Tesla a genius? Yes. Was he a visionary? Yes, and most of his biographers include discussions of his use of imagination in the early stages of his inventing process. Appropriately, Carlson quotes Pablo Picasso– "Everything you can imagine is real"–at the head of the chapter on Tesla's research into the wireless transmission of power. Another visionary, Mozart, was famous for editing his music in his head, committing it to paper only in final form, no more editing necessary. So too Tesla with many of his inventions.

Dr. Carlson's training was in science and technology, and we go to his book to get clear, technical explanations of how some of Tesla's most important inventions functioned. See, for example, the explanation of Tesla's discovery of the rotating magnetic field, which permitted him to eliminate those trouble-prone brushes on the DC motors of the day. And thanks to Carlson I can at last understand (sorta) how a *bladeless* turbine can be made to spin.

Tesla enthusiasts sometimes tout him as radio's true inventor. In 1897 he demonstrated a model of his unmanned, radio-controlled torpedo boat designed to carry bombs to the hulls of enemy warships. Tesla grandiosely claimed it could put an end to war. The boat, never fully realized, fascinated many people, including Mark Twain, who was keen to promote investment in it.

Biography allows readers to say, "Oh, if only he had . . ." over and over, and this is particularly true of Tesla. Examples: if only he had not relinquished his AC patents to Westinghouse. If only he had ditched the idea of wireless transmission of power. If only he had rightly interpreted one of his glass plate exposures a few months before Roentgen would claim the discovery of X-rays. If only he had brought some of his inventions to market instead of endlessly predicting stuff about them before they had been completed.

Think of Tesla the next time you turn on a fluorescent (he called it a phosphorescent) bulb. Here was yet another Tesla invention which he neglected to bring to finished, commercial form, too busy, perhaps, thinking or dreaming of his next innovation/discovery/invention.

Free-wheeling genius Tesla lived, thought and dreamed big. I am predicting there are more biographies to come.

Tesla circa 1894-95.
From Nikola Tesla Museum, Belgrade, Serbia.

Tube Lore II: A Reference for Users and Collectors
Ludwell Sibley. Beaver Press, 288 pages plus compact disc, 2019.
Pittsburgh Oscillator, v35 n1, March 2020, p. 13.

This is the second, enlarged edition of Sibley's classic *Tube Lore,* which first appeared in 1996. The book comes with a compact disc attached to the back cover containing a hundred-or-so pages of additional information about really esoteric stuff. I am tempted to write: if it's not in the book or the CD, you probably don't need it. Still, it's hard to believe this is only a small distillation of the thousands of pages of tube specs "out there somewhere."

Don't skip the 24 pages of front matter. Anyone who reads them is sure to learn something new. This 78-year-old reader learned three new words: *rebulbed, reheatered* and *snivet.* (Google it).

Equipment restorers will profit from a careful reading of "Tube Hospital: Repair & Reactivation." Most people give little thought to the possibility of repairing a tube but there are sometimes simple fixes to be made.

The bulk of the book consists of the "data chapters." Here is a typical entry:

6SN7GT( ) Heater 6.3 V @ 600 mA, μ 20, 7.5 W total diss. <8BD>. (RCA, 266, 3-3-41) {A4237} (AWV, NEC, RA, SY also). AKA (SigC) VT-231, (Br.) CV170 (Br. CV181, (Br. CV1988. ♥ 1943-48. Masses of them in "all" US radar synchronizers of WW II. Three found in "every " TV set as of, say, 1948-52. ENIAC computer, 1946, used 6550 of them. Forerunner of the ShE "Hi-Po 6S78" replacement. As of 1961, RA was apparently the only remaining source of plain 6SN7GTs, but also offered the "A" and "B." RCA converted its 6SN7GT to button stems. Withdrawn by RCA 1966 in favor of later versions. RCA filled later orders for JAN mil. 6SN7GTs by re-"etch"ing button-stem 6SN7GTBs, which were a better product. "Our 6SN7GT design using button stem will meet 6SN7GTA requirements."

Following the 6SN7's five lines of hard-core specs comes other info that only someone thoroughly steeped in tube experience would think to add–the lore of tube lore.

Another example from the lore section of everyone's favorite H-V rectifier, the 1B3GT:

> *Had silver type numbers on side until 1947;*
> *these were found to contribute to electrolysis*
> *of glass, so number was moved to base.*

My one gripe about the book: it needs a comprehensive index from tube type numbers to page numbers. Entries in the book are classified, and one doesn't always know under which classification heading to look for a particular tube. So, looking for info on the 6SN7GT? That's no problem. Just look under "Receiving Types Designated Under RMA-EIA System."

But I wanted to read about the beautiful 4J50 cavity magnetron pictured on the book's front cover. So I went to the classification "Military Types," but no 4J50 was listed. Turns out, that information was on the compact disc under the heading "1A21 Weird". To be sure, Sibley does warn us in the book's foreword, "There is no simple way to organize a guide like this." Okay, enough complaining.

All things tubic

There is a staggering amount of information in this book and CD and it's hard to believe that just one man compiled it. Anyone with a casual interest–on up through obsession–will find the collection useful or just plain interesting. As the saying goes, once picked up the book is hard to put down. It is a remarkable achievement.

Heathkit Manual 1300 Full Technician Manuals Archive. On one flash drive. Available from *ebay.com,* search for *Heathkit Manual 1300.*
Pittsburgh Oscillator, v36 n1, March 2021, p. 8.

I haven't looked at all 1,293 files on this flash drive, but the ones I did see were impressive. Schematics, text and other data showed good resolution on my monitor and equally good resolution when I printed them on my laser printer.

These files obviously have many uses. I immediately went to the manual for Heath's T-3 signal tracer. I acquired a T-3 years ago at a yard sale, but it came with a broken RF probe. The clear instructions in the manual were useful in doing a rebuild.

A principle of journalism obligates me to report that pages three through eight of the T-3 manual have been omitted on the flash drive copy. I am hoping that this is a rare anomaly. As a check, I looked at ten other files at random and did page counts. No other anomalies were found.

Return with me now to the days of yesteryear, when an ordinary day was suddenly made special by the arrival in our mailbox of the latest Heath catalog. As a value-added feature, the flash drive includes pdf's for Heath catalogs of 1947, '48, '51, '58, '63, '66 '67, '68 '69, '71 (118 pages!), '76 and '78. Warning–these catalogs all exude nostalgia, and you may find yourself spending as much time with them as with the manuals.

More unexpected value-added files include the *Heathkit Master Parts List*, (65 pages), the *Heathkit Components Cross Reference*, (54 pages) and the *Tube Substitution Handbook* by the Howard W. Sams Engineering Staff, (128 pages, 21st edition, 1980). This last can be hard to find; look for "Tube Sub 1980".

It has been said of many books that once picked up they are hard to put down. So too with this flash drive. Once you've installed it in your computer, it will be difficult to pull it out of the USB port *mutatis mutandis*. It is highly recommended–with the obvious caveat that not all 1,293 files were viewed.

Despite Heath's recent bumpy history, the restructured company is back in the kit game. See their latest products at *www.heathkit.com*

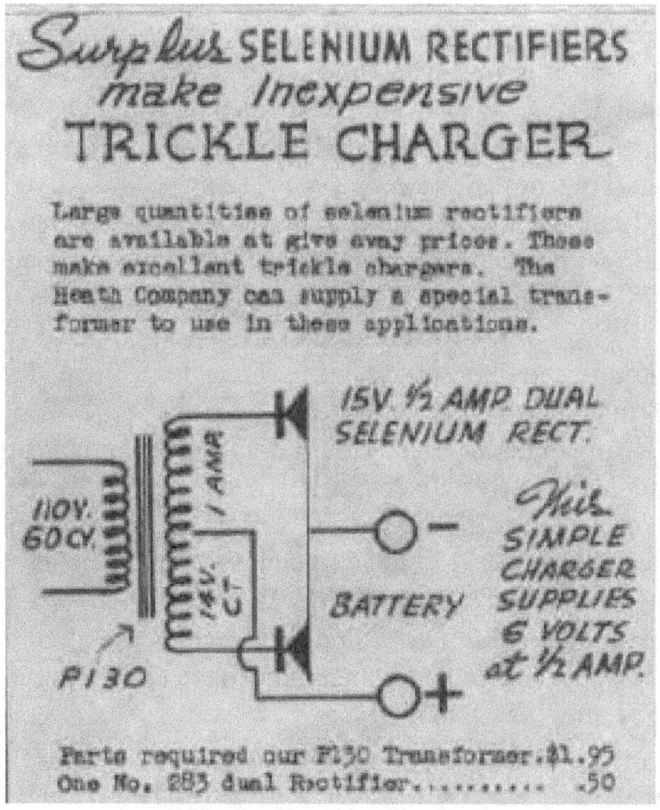

From the 1947 Heath catalog.

Raymond or Life and Death
Sir Oliver Lodge, FRS, George H. Doran Company, 1916, 404
pages.
Pittsburgh Oscillator, v36 n2, June 2021, p. 13.

During his long life (1851-1940) British physicist and psychic Sir
Oliver J. Lodge pursued many scientific inquiries. For years he
conducted dangerous experiments designed to prove the existence (or
non existence) of æther, the supposed carrier of electromagnetic
waves. Einstein would later use Lodge's findings in the formation of
his famous theories.

Lodge also built physical models to help explain to the public the
abstruse meanings of James Clerk Maxwell's theory of electromagnetism.

In 1894 Lodge demonstrated before the British Association the
wireless transmission of intelligence. His U.S. patent 609,154 of
1898 showed the advantage of using tuned radio circuits. Marconi's
later patent did not. In 1911 the Marconi company purchased rights
to Lodge's patent.

By 1935 the list of publications by Lodge numbered 1,154,
including 40 books.[1] More than 200 of his articles appeared in the
prestigious science journal *Nature*.

A different view of genius

So Lodge certainly had scientific *bona fides* in spades but his
stature may have been tarnished somewhat by his interest in
spiritualism, which in Victorian times meant communication with
the dead, particularly with dead relatives. (Here Lodge certainly had
ample opportunity; his father had 24 brothers and sisters.)[2]

On September 14, 1915 Lodge's youngest son Raymond was killed
at the front in Flanders in World War One. In many pages of
Raymond or Life and Death Lodge describes in minute detail the
attempts he and other family members made to communicate with
Raymond after his death.

Seances, sittings, mediums

At seances (Lodge called them "sittings") he and his family tried
to communicate with Raymond through mediums, people who were

supposed to be able to communicate with the dead. Sometimes the seances were conducted without a medium present. Sometimes they involved tables tilting on their own accord and even making sounds. One tilting from a table was interpreted as a *No* from Raymond, three tiltings in a row, a *Yes*. Lodge gave limited credence to these occurrences; nevertheless he and his family devised test questions for Raymond, questions to which only they and Raymond could have known the answers.

Lodge put himself and his beliefs "under the microscope" in this book. Ever the scientist, he did not prevaricate. He stated: "I am as convinced of continued existence, on the other side of death, as I am of existence here" (page 375) and, this time metaphorically, "The evidence [for continued existence beyond death] is cumulative, and has broken the back of all legitimate and reasonable scepticism" (page 288). Despite, or perhaps because of, the unusual subject of *Raymond or Life and Death*, the book became what we would today call a best-seller, due in part, no doubt, to the grief of the many who lost sons in the War.

Did Lodge's interest in spiritualism compromise his stellar reputation in the world of radio and science? Can living people communicate with dead people? You decide.

1. Besterman, Theodore, *Bibliography of Sir Oliver Lodge F.R.S.,* Oxford University Press, 1935.
2. *The OTB*, volume 43, number 3, August 2002, pp. 53-54.

On Air – the Broadcast
A Radio Serial Musical. Music by Matt Connor; book and lyrics by Stephen Gregory Smith.
Produced by Creative Cauldron, Duquesne University, and La Ti Do Productions, 2019, 2020. One hour and 34 minutes.

A *Pittsburgh Oscillator* Extra, v35 n4, December 2020, insert.

Because you're reading this we now know you're interested in radio history. If you are also interested in musicals, you've come to the right place.

The premiere of *On Air* as a stage musical was at Falls Church, Virginia in 2019. It was to be performed again at Duquesne University this year, but Covid-19 put an end to those plans. Not letting a mere pandemic get in their way, the producers have now made available an adaptation of the musical that could be called *On Air Lite*, playing at your nearest internet terminal.

The new production features the songs from the original, plus still photos apparently from the 2019 show. Tickets are available now at:
https://www.creativecauldron.org/on-air-broadcast.html

The tickets are free, "though your support is appreciated"–and richly deserved in this reviewer's opinion. Your ticket will link to the appropriate YouTube address. Note that the production is available online anytime, but only until December 13, so don't hesitate.

The songs amount to vignettes, sometimes sad, mostly funny, always imaginative. They portray life in and around the Frank and Flora Conrad house just before, during and after the birth of KDKA on November 2nd, 1920. Westinghouse vice president H. P. Davis and his wife Agnes are also present.

So we have musical portrayals of Flora Conrad and Agnes Davis commiserating with each other by telephone over the frequent extended absences of their respective husbands because of pressures from work.

Or Frank Conrad's attempts to get his young son Francis to "think a little bit further now."

Or Francis's own whimsical attempts to advance his budding career as a Health Department officer.

Another song imagines what happened when Frank and Flora took to dancing to the gramophone record playing in Frank's garage workshop while his experimental transmitter was turned on.

Also depicted in song is the chaos that ensued in the Conrad house on the night of the famous broadcast announcing the winner of the Harding-Cox presidential race.

(Note to producers: the elderly among us with compromised hearing abilities would have appreciated a chyron showing the lyrics.)

Garrett Matthews (Harry P. Davis), Erin Granfield (Agnes Davis), Robert Aubry Davis (Radio voice), Nora Palka (Flora Conrad) and Jimmy Mavrikes (Frank Conrad) are to be congratulated for this first-rate achievement. The voice of Pittsburgh radio personality Jack Bogut is also heard, introducing various segments of the production.

PARS members (and others) should not miss this charming, lively musical. Be sure to tell your friends.

> *Here is imaginative Pittsburgh radio history presented in a way never before experienced by PARS members.*

Book Availability

Chances are pretty good that your local bookstore will not have the book you discovered here and decided to buy. But chances are also pretty good that the book will be available at:

www.amazon.com

Other sources to try are:

www.abebooks.com

www.alibris.com

www.lulu.com

Take really tough cases to:

www.bookfinder.com

To find libraries that hold a particular book, go to:

www.worldcat.org

Author Index

David Kraeuter with 1935 Crosley Fiver Deluxe.

Pinhole paper negative, silver gelatin contact print. Copyright © 2005 by Allen C. Benson. Reprinted with permission. All rights reserved.

Other books by the author:

Radio and Television Pioneers: A Patent Bibliography, Scarecrow Press, 1992.

British Radio and Television Pioneers: A Patent Bibliography, Scarecrow Press, 1993.

Radio and Electronics Pioneers: A Patent Bibliography, UMI, 1994.

Index to Radio and Electronics Patents, UMI, 1995.

Numerical Index to Radio and Electronics Patents, UMI, 1997.

Radio Patent Lists and Index, 1830-1980, Edwin Mellen Press, 2001.

Ten Patents from Radio History, www.lulu.com, 2007.

Electronic Reviews: Hundreds of Thoughts on 100 Books, www.lulu.com, 2008.

A Radio Patent Chronology, www.lulu.com, 2009.

Electronic Essays, 6th edition, www.lulu.com, 2012.

1 2 3 4 5 6: A David Kraeuter Sampler, www.lulu.com, 2017.

The At Home Series, www.lulu.com, 2018

———————————————